Fundamentals of Proteomics

Fundamentals of Proteomics

George Wright

STATES
ACADEMIC PRESS
www.statesacademicpress.com

States Academic Press,
109 South 5th Street,
Brooklyn, NY 11249, USA

Visit us on the World Wide Web at:
www.statesacademicpress.com

ISBN: 978-1-63989-231-0 (Hardback)

Cataloging-in-Publication Data

Fundamentals of proteomics / George Wright.
p. cm.
Includes bibliographical references and index.
ISBN 978-1-63989-231-0
1. Proteomics. 2. Proteins. 3. Molecular biology. I. Wright, George.
QP551 .F86 2022
572.6--dc23

Table of Contents

Preface

Proteome is the set of proteins which is produced by an organism or a system. The large scale study of these protein structures is termed as proteomics. It involves the application of mass spectrometry, protein fractionation techniques and bioinformatics. Proteomics has enabled the identification of a vast array of proteins. This differs with time and unique requirements, or stresses, that a cell or organism undergoes. The processing and analysis of proteomics data is a difficult process but can result in the discovery of new disease biomarkers for clinical and diagnostic applications. Proteomics is a multidisciplinary field which has benefitted greatly from the genetic information of various genome projects. The topics included in this book on proteomics are of utmost significance and bound to provide incredible insights to readers. The extensive content of this book provides the readers with a thorough understanding of the subject. Coherent flow of topics, student-friendly language and extensive use of examples make this book an invaluable source of knowledge.

To facilitate a deeper understanding of the contents of this book a short introduction of every chapter is written below:

Chapter 1- Proteomics is the large scale study of an entire collection of proteins produced by a cell, tissue or an organism. Its primary purpose is to understand the structure and function of a protein set. This chapter introduces the reader to proteomics and its different types in a clear and concise manner for a better understanding of the topic.

Chapter 2- Protein-protein interactions are physical contacts between proteins in the cell as a result of biochemical events. Hydrogen bonds, covalent bonds, non-covalent bonds, etc. are the various bonds which contribute in protein-protein interactions. This chapter touches on the mentioned topics in a detailed manner for the benefit of the reader.

Chapter 3- Amino acids are the building blocks of proteins held together by peptide bonds. The chemical cycle of breaking down proteins into amino acids and simple derivative compounds for the synthesizing of new molecules to be used in the body is called protein breaking. This chapter focuses on proteins and their break down, and protein structures to provide a clearer understanding of the subject matter.

Chapter 4- Cyanovirin-N is a protein that displays virucidal activity against several viruses. Blood proteins are present in blood plasma and are also known as plasma proteins. Globular proteins are water-soluble, spherical proteins which are fairly common, hemoglobin being one of them. This chapter delves into the various proteins studied in proteomics and covers the aforementioned proteins in a comprehensive manner.

Chapter 5- There are many methods and techniques used to investigate protein-protein interactions. A few of them are affinity chromatography, ChIP-sequencing, phosphotyrosine-binding domain, LIM domain, sterile alpha motif, FERM domain, protein subunit, RNA-binding protein, Pleckstrin homology domain, etc. The topics elaborated in this chapter will help in gaining a better perspective of this vast field.

Finally, I would like to thank the entire team involved in the inception of this book for their valuable time and contribution. This book would not have been possible without their efforts. I would also like to thank my friends and family for their constant support.

George Wright

Chapter 1

Proteomics: An Introduction

Proteomics is the large scale study of an entire collection of proteins produced by a cell, tissue or an organism. Its primary purpose is to understand the structure and function of a protein set. This chapter introduces the reader to proteomics and its different types in a clear and concise manner for a better understanding of the topic.

The term "proteomics" was first coined in 1995 and was defined as the large-scale characterization of the entire protein complement of a cell line, tissue, or organism. Today, two definitions of proteomics are encountered. The first is the more classical definition, restricting the large-scale analysis of gene products to studies involving only proteins. The second and more inclusive definition combines protein studies with analyses that have a genetic readout such as mRNA analysis, genomics, and the yeast two-hybrid analysis. However, the goal of proteomics remains the same, i.e., to obtain a more global and integrated view of biology by studying all the proteins of a cell rather than each one individually.

Using the more inclusive definition of proteomics, many different areas of study are now grouped under the rubric of proteomics. These include protein-protein interaction studies, protein modifications, protein function, and protein localization studies to name a few. The aim of proteomics is not only to identify all the proteins in a cell but also to create a complete three-dimensional (3-D) map of the cell indicating where proteins are located. These ambitious goals will certainly require the involvement of a large number of different disciplines such as molecular biology, biochemistry, and bioinformatics. It is likely that in bioinformatics alone; more powerful computers will have to be devised to organize the immense amount of information generated from these endeavors.

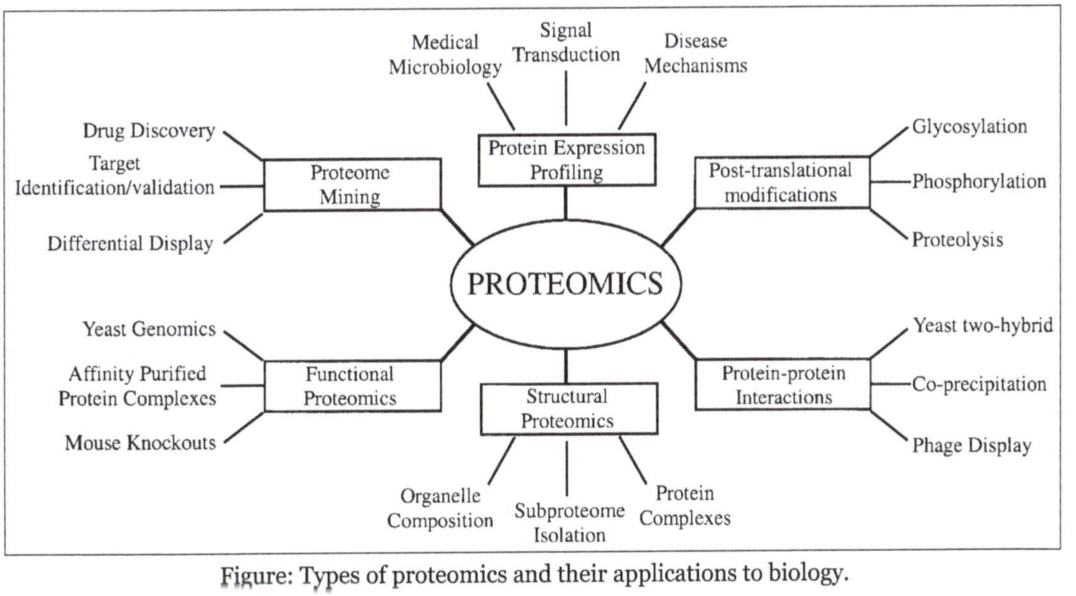

Figure: Types of proteomics and their applications to biology.

In the quest to characterize the proteome of a given cell or organism, it should be remembered that the proteome is dynamic. The proteome of a cell will reflect the immediate environment in which it is studied. In response to internal or external cues, proteins can be modified by posttranslational modifications, undergo translocations within the cell, or be synthesized or degraded. Thus, examination of the proteome of a cell is like taking a "snapshot" of the protein environment at any given time. Considering all the possibilities, it is likely that any given genome can potentially give rise to an infinite number of proteomes.

The first protein studies that can be called proteomics began in 1975 with the introduction of the two-dimensional gel by O'Farrell, Klose, and Scheele, who began mapping proteins from Escherichia coli, mouse, and guinea pig, respectively. Although many proteins could be separated and visualized, they could not be identified. Despite these limitations, shortly thereafter a large-scale analysis of all human proteins was proposed. The goal of this project, termed the human protein index, was to use two-dimensional protein electrophoresis (2-DE) and other methods to catalog all human proteins. However, lack of funding and technical limitations prevented this project from continuing.

Although the development of 2-DE was a major step forward, the science of proteomics would have to wait until the proteins displayed by 2-DE could be identified. One problem that had to be overcome was the lack of sensitive protein sequencing technology. Improving sensitivity was critical for success because biological samples are often limiting and both one-dimensional (1-D) and two-dimensional (2-D) gels have limits in protein loading capacity. The first major technology to emerge for the identification of proteins was the sequencing of proteins by Edman degradation. A major breakthrough was the development of microsequencing techniques for electroblotted proteins. This technique was used for the identification of proteins from 2-D gels to create the first 2-D databases. Improvements in microsequencing technology resulted in increased sensitivity of Edman sequencing in the 1990s to high-picomole amounts.

One of the most important developments in protein identification has been the development of MS technology. In the last decade, the sensitivity of analysis and accuracy of results for protein identification by MS have increased by several orders of magnitude. It is now estimated that proteins in the femtomolar range can be identified in gels. Because MS is more sensitive, can tolerate protein mixtures, and is amenable to high-throughput operations, it has essentially replaced Edman sequencing as the protein identification tool of choice.

Genome Information

The growth of proteomics is a direct result of advances made in large-scale nucleotide sequencing of expressed sequence tags and genomic DNA. Without this information, proteins could not be identified even with the improvements made in MS. Protein identification (by MS or Edman sequencing) relies on the presence of some form of database for the given organism. The majority of DNA and protein sequence information has accumulated within the last 5 to 10 years. In 1995, the first complete genome of an organism was sequenced, that of Haemophilus influenzae. At the time of this writing, the sequencing of the genomes of 45 microorganisms has been completed and that of 170 more is under way. To date, five eukaryotic genomes have been completed: Arabidopsis thaliana, Saccharomyces cerevisiae, Schizosaccharomyces pombe, Caenorhabditis elegans, and Drosophila melanogaster. In addition, the rice, mouse, and human genomes are near completion.

Need of Proteomics

Many types of information cannot be obtained from the study of genes alone. For example, proteins, not genes, are responsible for the phenotypes of cells. It is impossible to elucidate mechanisms of disease, aging, and effects of the environment solely by studying the genome. Only through the study of proteins can protein modifications be characterized and the targets of drugs identified.

Annotation of the Genome

One of the first applications of proteomics will be to identify the total number of genes in a given genome. This "functional annotation" of a genome is necessary because it is still difficult to predict genes accurately from genomic data. One problem is that the exon-intron structure of most genes cannot be accurately predicted by bioinformatics. To achieve this goal, genomic information will have to be integrated with data obtained from protein studies to confirm the existence of a particular gene.

Protein Expression Studies

In recent years, the analysis of mRNA expression by various methods has become increasingly popular. These methods include serial analysis of gene expression (SAGE) and DNA microarray technology. However, the analysis of mRNA is not a direct reflection of the protein content in the cell. Consequently, many studies have now shown a poor correlation between mRNA and protein expression level.

The formation of mRNA is only the first step in a long sequence of events resulting in the synthesis of a protein. First, mRNA is subject to posttranscriptional control in the form of alternative splicing, polyadenylation, and mRNA editing. Many different protein isoforms can be generated from a single gene at this step. Second, mRNA then can be subject to regulation at the level of protein translation. Proteins, having been formed, are subject to posttranslational modification. It is estimated that up to 200 different types of posttranslational protein modification exist. Proteins can also be regulated by proteolysis and compartmentalization. The average number of protein forms per gene was predicted to be one or two in bacteria, three in yeast, and three or more in humans. Therefore, it is clear that the tenet of "one gene, one protein" is an oversimplification. In addition, some bodily fluids such as serum or urine have no mRNA source and therefore cannot be studied by mRNA analysis.

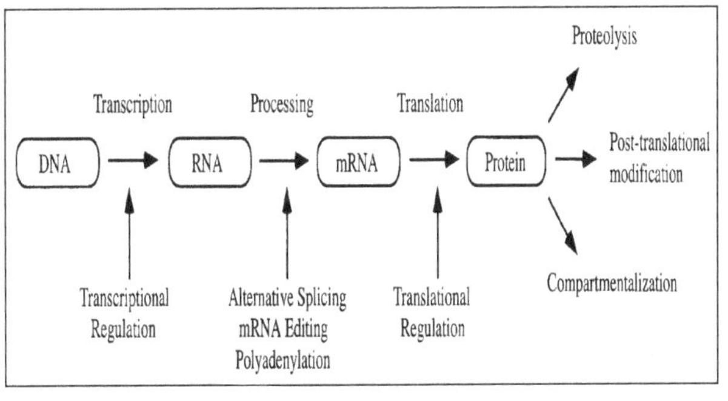

Figure shows Mechanisms by which a single gene can give rise to multiple gene products. Multiple pro-tein isoforms can be generated by RNA processing when RNA is alternatively spliced or edited to form mature mRNA. mRNA, in turn, can be regulated by stability and efficiency of translation. Proteins can be regulated by additional mechanisms, including posttranslational modification, proteolysis, or compartmentalization.

Protein Function

According to one study, no function can be assigned to about one-third of the sequences in organisms for which the genomes have been sequenced. The complete identification of all proteins in a genome will aid the field of structural genomics in which the ultimate goal is to obtain 3-D structures for all proteins in a proteome. This is necessary because the functions of many proteins can only be inferred by examination of their 3-D structure.

Protein Modifications

One of the most important applications of proteomics will be the characterization of posttranslational protein modifications. Proteins are known to be modified posttranslationally in response to a variety of intracellular and extracellular signals. For example, protein phosphorylation is an important signaling mechanism and disregulation of protein kinases or phosphatases can result in oncogenesis. By using a proteomics approach, changes in the modifications of many proteins expressed by a cell can be analyzed simultaneously.

Protein Localization and Compartmentalization

One of the most important regulatory mechanisms known is protein localization. The mislocalization of proteins is known to have profound effects on cellular function (e.g., cystic fibrosis). Proteomics aims to identify the subcellular location of each protein. This information can be used to create a 3-D protein map of the cell, providing novel information about protein regulation.

Protein-Protein Interactions

Of fundamental importance in biology is the understanding of protein-protein interactions. The process of cell growth, programmed cell death, and the decision to proceed through the cell cycle are all regulated by signal transduction through protein complexes. Proteomics aims to develop a complete 3-D map of all protein interactions in the cell. One step toward this goal was recently completed for the microorganism Helicobacter pylori. Using the yeast two-hybrid method to detect protein interactions, 1,200 connections were identified between H. pylori proteins covering 46.6% of the genome. A comprehensive two-hybrid analysis has also been performed on all the proteins from the yeast S. cerevisiae.

Types of Proteomics

Protein Expression Proteomics

The quantitative study of protein expression between samples that differ by some variable is known as expression proteomics. In this approach, protein expression of the entire proteome or of subproteomes between samples can be compared. Information from this approach can identify novel proteins in signal transduction or identify disease-specific proteins.

Structural Proteomics

Proteomics studies whose goal is to map out the structure of protein complexes or the proteins present in a specific cellular organelle are known as "cell map" or structural proteomics. Structural proteomics attempts to identify all the proteins within a protein complex or organelle, determine where they are located, and characterize all protein-protein interactions. An example of structural proteomics was the recent analysis of the nuclear pore complex. Isolation of specific subcellular organelles or protein complexes by purification can greatly simplify the proteomic analysis. This information will help piece together the overall architecture of cells and explain how expression of certain proteins gives a cell its unique characteristics.

Functional Proteomics

"Functional proteomics" is a broad term for many specific, directed proteomics approaches. In some cases, specific subproteomes are isolated by affinity chromatography for further analysis. This could include the isolation of protein complexes or the use of protein ligands to isolate specific types of proteins. This approach allows a selected group of proteins to be studied and characterized and can provide important information about protein signaling, disease mechanisms or protein-drug interactions.

Technology of Proteomics

An integral part of the growth of proteomics has been in the advances made in protein technologies. Twenty-six years ago, when 2-DE was introduced, very few tools existed for proteomics. Since that time, new technologies have emerged and old ones have been improved in areas from protein separation to protein identification. However, it is also clear that it is still not feasible to conduct many types of proteomics because of limitations in technology. These problems will have to be solved and new technologies must be developed for proteomics to reach its full potential. A typical proteomics experiment (such as protein expression profiling) can be broken down into the following categories: (i) the separation and isolation of proteins from a cell line, tissue, or organism; (ii) the acquisition of protein structural information for the purposes of protein identification and characterization; and (iii) database utilization.

Separation and Isolation of Proteins

By the very definition of proteomics, it is inevitable that complex protein mixtures will be encountered. Therefore, methods must exist to resolve these protein mixtures into their individual components so that the proteins can be visualized, identified, and characterized. The predominant technology for protein separation and isolation is polyacrylamide gel electrophoresis. Unlike the breakthroughs in molecular biology that eventually enabled the sequencing of the human genome, some aspects of protein science have shown little progress over the years. Protein separation technology is one of them. Since its inception some 32 years ago, protein electrophoresis still remains the most effective way to resolve a complex mixture of proteins. In many applications, it is at this stage where the bottleneck occurs. This is because 1- or 2-DE is a slow, tedious procedure that is not easily automated. However, until something replaces this methodology, it will remain an essential component of proteomics.

One- and Two-Dimensional Gel Electrophoresis

For many proteomics applications, 1-DE is the method of choice to resolve protein mixtures. In 1-DE, proteins are separated on the basis of molecular mass. Because proteins are solubilized in sodium dodecyl sulfate (SDS), protein solubility is rarely a problem. Moreover, 1-DE is simple to perform, is reproducible, and can be used to resolve proteins with molecular masses 10 to 300 kDa. The most common application of 1-DE is the characterization of proteins after some form of protein purification. This is because of the limited resolving power of a 1-D gel. If a more complex protein mixture such as a crude cell lysate is encountered, then 2-DE can be used. In 2-DE, proteins are separated by two distinct properties. They are resolved according to their net charge in the first dimension and according to their molecular mass in the second dimension. The combination of these two techniques produces resolution far exceeding that obtained in 1-DE.

One of the greatest strengths of 2-DE is the ability to resolve proteins that have undergone some form of posttranslational modification. This resolution is possible in 2-DE because many types of protein modifications confer a difference in charge as well as a change in mass on the protein. One such example is protein phosphorylation. Frequently, the phosphorylated form of a protein can be resolved from the nonphosphorylated form by 2-DE. In this case, a single phosphoprotein will appear as multiple spots on a 2-D gel. In addition, 2-DE can detect different forms of proteins that arise from alternative mRNA splicing or proteolytic processing.

The primary application of 2-DE continues to be protein expression profiling. In this approach, the protein expression of any two samples can be qualitatively and quantitatively compared. The appearance or disappearance of spots can provide information about differential protein expression, while the intensity of those spots provides quantitative information about protein expression levels. Protein expression profiling can be used for samples from whole organisms, cell lines, tissues, or bodily fluids. Examples of this technique include the comparison of normal and diseased tissues or of cells treated with various drugs or stimuli. An example of 2-DE used in protein profiling is shown in Figure below.

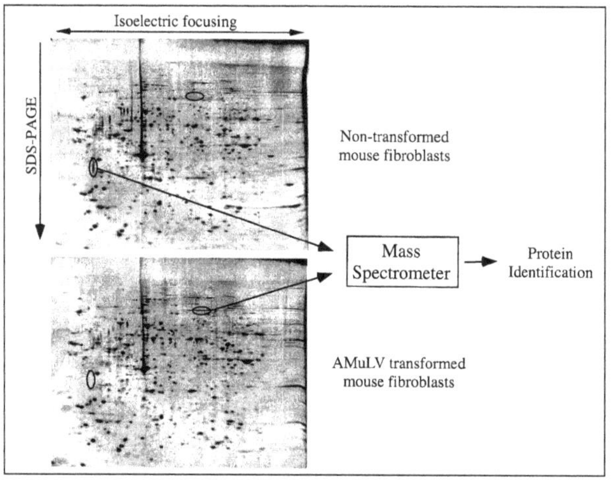

Figure shows Protein expression profiling by 2-DE. Whole-cell lysates from nontransformed and Abel-son murine leukemia virus (AMuLV)- transformed mouse fibroblasts were resolved by 2-DE, and proteins were visualized by silver staining. Differentially expressed proteins were excised from the gel and identified by MS.

Another application of 2-DE is in cell map proteomics. 2-DE is used to map proteins from microorganisms, cellular organelles, and protein complexes. It can also be used to resolve and characterize proteins in subproteomes that have been created by some form of purification of a proteome. Because a single 2-DE gel can resolve thousands of proteins, it remains a powerful tool for the cataloging of proteins. Many 2-DE databases have been constructed and are available on the World Wide Web.

A number of improvements have been made in 2-DE over the years. One of the biggest improvements was the introduction of immobilized pH gradients, which greatly improved the reproducibility of 2-DE. The use of fluorescent dyes has improved the sensitivity of protein detection, and specialized pH gradients are able to resolve more proteins. The speed of running 2-DE has been improved, and 2-D gels can now be run in the minigel format. In addition, there have been efforts to automate 2-DE. Hochstrasser's group has automated the process of 2-DE from gel running to image analysis and spot picking. The use of computers has aided the analysis of complex 2-D gel images. This is a critical aspect of 2-DE because a high degree of accuracy is required in spot detection and annotation if artifacts are to be avoided. Recently, a molecular scanner was developed to record 2-DE images. Software programs such as Melanie compare computer images of 2-D gels and facilitate both the identification and quantitation of protein spots between samples. A recent exciting advance in 2-DE was developed by Minden and coworkers. This technology is called difference gel electrophoresis (DIGE) and utilizes fluorescent tagging of two protein samples with two different dyes. The tagged proteins are run on the same 2-D gel, and postrun fluorescence imaging of the gel is used to create two images, which are superimposed to identify pattern differences. The dyes are amine reactive and are designed to ensure that proteins common to both samples have the same relative mobility regardless of the dye used to tag them. This technique circumvents the need to compare several 2-D gels. DIGE was used to detect differences between exogenous proteins in two D. melanogaster embryo extracts at nanogram levels. Moreover, an inducible protein from Escherichia coli was detected after 15 min of induction. This technology is now commercially available from Amersham/Pharmacia.

However, a number of problems with 2-DE still remain. Despite efforts to automate protein analysis by 2-DE, it is still a labor-intensive and time-consuming process. A typical 2-DE experiment can take two days, and only a single sample can be analyzed per gel. In addition, 2-DE is limited by both the number and type of proteins that can be resolved. For example, the protein mixture obtained from a eukaryotic cell lysate is too complex to be completely resolved on a single 2-D gel. Many large or hydrophobic proteins will not enter the gel during the first dimension, and proteins of extreme acidity or basicity (proteins with pIs below pH 3 and above pH 10) are not well represented. Some of these problems can be overcome with different solubilization conditions and pH gradients. Another limitation of 2-DE is the inability to detect low-copy proteins when a total-cell lysate is analyzed. In a crude cell extract, the most abundant proteins can dominate the gel, making the detection of low-copy proteins difficult. It was determined in the analysis of yeast proteins by 2-DE that no proteins defined as low-copy proteins were visible by 2-DE. Yet it is estimated that over half of the 6,000 genes in yeast may encode low-copy proteins. In mammalian cells, the dynamic range of protein expression is estimated to be between 7 and 9 orders of magnitude. This problem cannot be overcome by simply loading more protein on the gel, because the resolution will decrease and the comigration of proteins will increase. Because of these limitations, the largest application of 2-DE in the future will probably involve the analysis of protein complexes or subproteomes as opposed to whole proteomes.

Alternatives to Electrophoresis

The limitations of 2-DE have inspired a number of approaches to bypass protein gel electrophoresis. One approach is to convert an entire protein mixture to peptides (usually by digestion with trypsin) and then purify the peptides before subjecting them to analysis by MS. Various methods for peptide purification have been devised, including liquid chromatography, capillary electrophoresis, and a combination of techniques such as multidimensional protein identification or cation-exchange chromatography and reverse-phase (RP) chromatography. The advantage of these methods is that because a 2-D gel is avoided, a greater number of proteins in the mixture can be represented. The disadvantage is that it can require an immense amount of time and computing power to deconvolute the data obtained. In addition, considerable time and effort may be expended in the analysis of uninteresting proteins. One of the most exciting techniques to emerge as an alternative to protein electrophoresis is that of isotope-coded affinity tags (ICAT). This method allows the quantitative protein profiling between different samples without the use of electrophoresis.

Acquisition of Protein Structure Information

Edman Sequencing

One of the earliest methods used for protein identification was microsequencing by Edman chemistry to obtain N-terminal amino acid sequences. Little has changed in Edman chemistry since its introduction, but improvements in sequencing technology have increased the sensitivity and ease of Edman sequencing. Although the use of Edman sequencing is waning in the field of proteomics, it is still a very useful tool for several reasons. First, because Edman sequencing existed before MS as a sequencing tool, a considerable number of investigators continue to use Edman sequencing. Second, Edman sequencing of relatively abundant proteins is a viable alternative to MS if a mass spectrometer is in high demand for the identification of low-copy proteins or is not available. Finally, Edman sequencing is used to obtain the N-terminal sequence of a protein (if possible) to determine its true start.

The N-terminal sequencing of proteins was introduced by Edman in 1949. Today, Edman sequencing is most often used to identify proteins after they are transferred to membranes. The development of membranes compatible with sequencing chemicals allowed Edman sequencing to become a more applicable sequencing method for the identification of proteins separated by SDS-polyacrylamide gel electrophoresis. One of the biggest problems that has limited the success of Edman sequencing in the past is N-terminal modification of proteins. Since it is difficult to tell if a protein is N-terminally blocked before it is sequenced, precious samples were often lost in failed sequencing attempts. To overcome this problem, we developed a novel approach called mixed-peptide sequencing. In mixed-peptide sequencing, a protein is converted into peptides by cleavage with cyanogen bromide (CNBr) or skatole and the peptides are sequenced in an Edman sequencer simultaneously.

Briefly, the process of mixed-peptide sequencing involves separation of a complex protein mixture by polyacrylamide gel electrophoresis (1-D or 2-D) and then transfer of the proteins to an inert membrane by electroblotting. The proteins of interest are visualized on the membrane surface, excised, and fragmented chemically at methionine (by CNBr) or tryptophan (by skatole) into several large peptide fragments. On average, three to five peptide fragments are generated, consistent with the frequency of occurrence of methionine and tryptophan in most proteins. The membrane piece is placed directly into an automated Edman sequencer without further manipulation. Between 6 and 12 automated Edman cycles are carried out (4 to 8 h), and the mixed-sequence data are fed into the FASTF or TFASTF algorithms, which sort and match the data against protein (FASTF) and DNA (TFASTF) databases to unambiguously identify the protein. The FASTF and TFASTF programs were written in collaboration with William Pearson.

Because minimal sample handling is involved, mixed-peptide sequencing can be a sensitive approach for identifying proteins in polyacrylamide gels at the 0.1- to 1-pmol level. The mixed sequence approach has the advantage of enabling subsequent searches to be carried out against unannotated or non-species specific DNA databases as well as annotated protein databases. This is because the T/FASTF algorithms utilize actual amino acid sequence and are therefore able to tolerate errors in the database as well as polymorphisms or conservative substitutions. A recent variation of T/FASTF has been devised for MS.

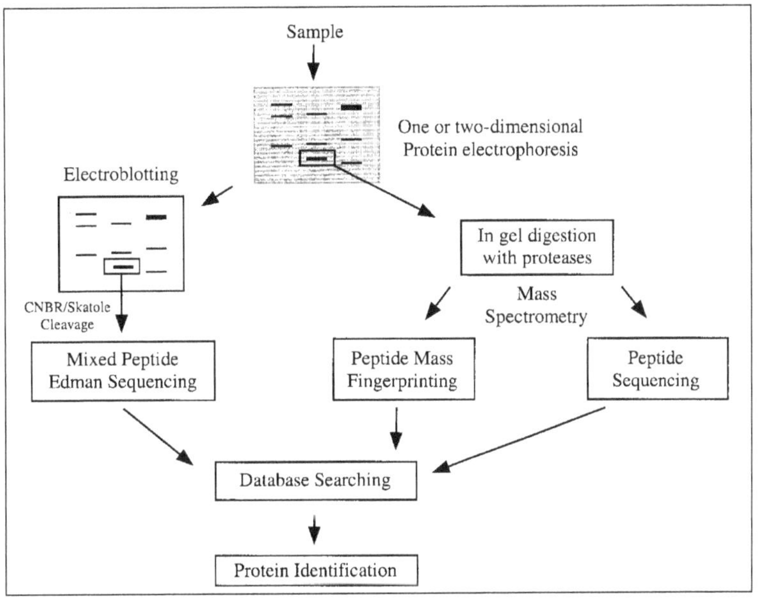

Figure: Strategies for protein identification. The identification of proteins from a polyacrylamide gel by mixed-peptide sequencing or MS is depicted. For mixed-peptide sequencing, proteins are transferred to a membrane and cleaved with CNB or skatole, and the resulting peptides are sequenced simultaneously by Edman degradation. For MS, proteins are in-gel digested with proteases and the resulting peptides are mass fingerprinted or sequenced. Information from all these methods is used to search nucleotide and protein databases for protein identification.

Mass Spectrometry

MS enables protein structural information, such as peptide masses or amino acid sequences, to be obtained. This information can be used to identify the protein by searching nucleotide and protein databases. It also can be used to determine the type and location of protein modifications. The harvesting of protein information by MS can be divided into three stages: (i) Sample preparation, (ii) Sample ionization, and (iii) Mass analysis.

Sample Preparation

In most of proteomics, a protein is resolved from a mixture by using a 1- or 2-D polyacrylamide gel. The challenge is to extract the protein or its constituent peptides from the gel, purify the sample, and analyze it by MS. The extraction of whole proteins from gels is inefficient; however, if a protein is "in-gel" digested with a protease, many of the peptides can be extracted from the gel. A method for in-gel protein digestion was developed and is now commonly applied to both 1- and

2-D gels. In-gel digestion is more efficient at sample recovery than other common methods such as electroblotting. In addition, the conversion of a protein into its constituent peptides provides more information than can be obtained from the whole protein itself. For many applications, the peptides recovered following in-gel digestion need to be purified to remove gel contaminants. Common impurities from electrophoresis such as salts, buffers, and detergents can interfere with MS. In addition, peptide samples often require concentration before being analyzed by MS. One method of peptide purification commonly employed for this purpose is reverse-phase chromatography, which is available in a variety of formats. Peptides can be purified with ZipTips (Millipore) or Poros R2 perfusion material (PerSeptive Biosystems, Framingham, Mass.) or by high-pressure liquid chromatography (HPLC).

Sample Ionization

For biological samples to be analyzed by MS, the molecules must be charged and dry. This is accomplished by converting them to desolvated ions. The two most common methods for this are electrospray ionization (ESI) and matrix-assisted laser desorption/ionization (MALDI). In both methods, peptides are converted to ions by the addition or loss of one or more protons. ESI and MALDI are "soft" ionization methods that allow the formation of ions without significant loss of sample integrity. This is important because it enables accurate mass information to be obtained about proteins and peptides in their native states.

(a) Electrospray ionization: In ESI, a liquid sample flows from a microcapillary tube into the orifice of the mass spectrometer, where a potential difference between the capillary and the inlet to the mass spectrometer results in the generation of a fine mist of charged droplets. As the solvent evaporates, the sizes of the droplets decrease, resulting in the formation of desolvated ions. A significant improvement in ESI technology occurred with the development of nanospray ionization. In nanospray ionization, the microcapillary tube has a spraying orifice of 1 to 2 m and flow rates as low as 5 to 10 nl/min. The low flow rates possible with nanospray ionization reduce the amount of sample consumed and increase the time available for analysis. For ESI, there are several ways to deliver the sample to the mass spectrometer. The simplest method is to load individual microcapillary tubes with sample. Because a new microcapillary tube is used for each sample, cross-contamination is avoided. In ESI, peptides require some form of purification after in-gel digestion, and this can be accomplished directly in the microcapillary tubes. The drawback to both the purification and manual loading of microcapillary tubes is that it is tedious and slow. As an alternative, electrospray sources have been connected in line with liquid chromatography (LC) systems that automatically purify and deliver the sample to the mass spectrometer. Examples of this method are LC, reversephase LC (RP-LC) and reverse-phase microcapillary LC (RP-LC).

(b) Matrix-assisted laser desorption/ionization: In MALDI, the sample is incorporated into matrix molecules and then subjected to irradiation by a laser. The laser promotes the formation of molecular ions. The matrix is typically a small energy-absorbing molecule such as 2,5-dihydroxybenzoic acid or -cyano-4-hydroxycinnamic acid. The analyte is spotted, along with the matrix, on a metal plate and allowed to evaporate, resulting in the formation of crystals. The plate, which can be 96-well format, is then placed in the mass spectrometer, and the laser is automatically targeted to specific places on the plate. Since sample application can be performed by a robot, the entire process including data collection and analysis can be automated. This is the single biggest advantage of

MALDI. Another advantage of MALDI over ESI is that samples can often be used directly without any purification after in-gel digestion.

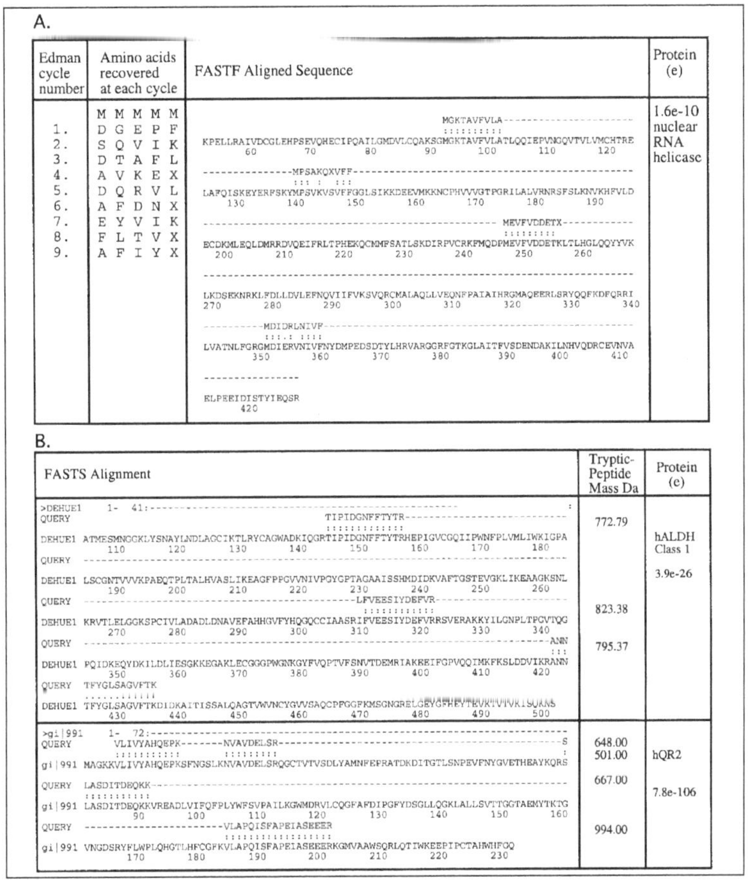

Figure: The FASTF and FASTS search programs.

(A) Example of a FASTF search where the amino acid sequence is obtained by Edman sequencing of a mixture of peptides. The information is then deconvoluted by a computer algorithm, and the results are given an expectation score (e). (B) With the FASTS program, a similar type of search is conducted except that peptide sequences obtained from MS are used.

Mass Analysis

Mass analysis follows the conversion of proteins or peptides to molecular ions. This is accomplished by the mass analyzers in a mass spectrometer, which resolve the molecular ions on the basis of their mass and charge in a vacuum.

(a) Quadrupole mass analyzers: One of the most common mass analyzers is the quadrupole mass analyzer. Here, ions are transmitted through an electric field created by an array of four parallel metal rods, the quadrupole. A quadrupole can act to transmit all ions or as a mass filter to allow the transmission of ions of a certain mass-to-charge (m/z) ratio. If multiple quadrupoles are combined, they can be used to obtain information about the amino acid sequence of a peptide.

(b) Time of flight: A time-of-flight (TOF) instrument is one of the simplest mass analyzers. It measures the m/z ratio of an ion by determining the time required for it to traverse the length of a flight tube. Some TOF mass analyzers include an ion mirror at the end of the flight tube, which reflects ions back through the flight tube to a detector. In this way, the ion mirror serves to increase the length of the flight tube. The ion mirror also corrects for small energy differences among ions. Both of these factors contribute to an increase in mass resolution.

(c) Ion trap: Ion trap mass analyzers function to trap molecular ions in a 3-D electric field. In contrast to a quadrupole mass analyzer, in which ions are discarded before the analysis begins, the main advantage of an ion trap mass analyzer is the ability to allow ions to be "stored" and then selectively ejected from the ion trap, increasing sensitivity.

Types of Mass Spectrometers

Most mass spectrometers consist of four basic elements:

- An ionization source,
- One or more mass analyzers,
- An ion mirror,
- A detector.

The names of the various instruments are derived from the name of their ionization source and the mass analyzer. The analysis of proteins or peptides by MS can be divided into two general categories:

- Peptide mass analysis,
- Amino acid sequencing.

In peptide mass analysis or peptide mass fingerprinting, the masses of individual peptides in a mixture are measured and used to create a mass spectrum. In amino acid sequencing, a procedure known as tandem mass spectrometry, or MS/MS, is used to fragment a specific peptide into smaller peptides, which can then be used to deduce the amino acid sequence.

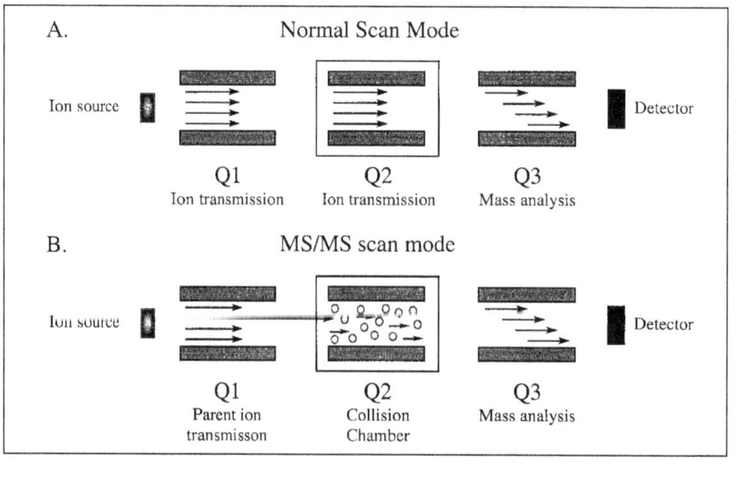

Figure shows MS/MS. Conventional and MS/MS modes of analysis in a triple-quadrupole mass spectrometer are shown. (A) In the normal scanning mode, all ions of a certain m/z range are transmitted through the first two quadrupoles for mass analysis in the third quadrupole. From this MS spectrum, a parent ion is selected for fragmentation in the collision cell. (B) In MS/MS mode, the parent ion is selectively transmitted into the collision chamber and fragmented, and the result-ing daughter ions are resolved in the third quadrupole.

(a) Triple quadrupole: Triple-quadrupole mass spectrometers are most commonly used to obtain amino acid sequences. In the first stage of analysis, the machine is operated in MS scan mode and all ions above a certain m/z ratio are transmitted to the third quadrupole for mass analysis. In the second stage, the mass spectrometer is operated in MS/MS mode and a particular peptide ion is selectively passed into the collision chamber. Inside the collision chamber, peptide ions are fragmented by interactions with an inert gas by a process known as collision-induced dissociation or collisionally activated dissociation. The peptide ion fragments are then resolved on the basis of their m/z ratio by the third quadrupole. Since two different mass spectra are obtained in this analysis, it is referred to as tandem mass spectrometry (MS/MS). MS/MS is used to obtain the amino acid sequence of peptides by generating a series of peptides that differ in mass by a single amino acid.

(b) Quadrupole-TOF: In recent years, several "hybrid" mass spectrometers have emerged from the combination of different ionization sources with mass analyzers. One example is the quadrupole-TOF mass spectrometer. In this machine, the first quadrupole (Q1) and the quadrupole collision cell (q) of a triple-quadrupole machine have been combined with a time-of-flight analyzer (TOF). The main applications of a QqTOF mass spectrometer are protein identification by amino acid sequencing and characterization of protein modifications. However, because it is coupled to electrospray, it is not typically utilized for large-scale proteomics.

(c) MALDI-TOF: The principal application of a MALDITOF mass spectrometer is peptide mass fingerprinting because it can be completely automated, making it the method of choice for large-scale proteomics work. Because of its speed, MALDI-TOF is frequently used as a first-pass instrument for protein identification. If proteins cannot be identified by fingerprinting, they can then be analyzed by electrospray and MS/MS. A MALDI-TOF machine can also be used to obtain the amino acid sequence of peptides by a method known as post-source decay. However, peptide sequencing by post-source decay is not as reliable as sequencing with competing electrospray methods because the peptide fragmentation patterns are much less predictable.

(d) MALDI-QqTOF: The MALDI-QqTOF mass spectrometer was developed to permit both peptide mass fingerprinting and amino acid sequencing. It was formed by the combination of a MALDI ion source with a QqTOF mass analyzer. Thus, if a sample is not identified by peptide mass fingerprinting in the first step, the amino acid sequence can then be obtained without having to use a different mass spectrometer. However, in our experience, the amino acid sequence information obtained using this instrument was more difficult to interpret than that obtained from a nanospray-QqTOF mass spectrometer.

(e) FT-ICR: A Fourier transform ion cyclotron resonance (FT-ICR) mass spectrometer is an ion-trapping instrument that can achieve higher mass resolution and mass accuracy than any other type of mass spectrometer. Recently, FTICR has been employed in the analysis of biomolecules ionized by both ESI and MALDI. The unique abilities of FT-ICR provide certain advantages compared to other mass spectrometers. For example, because of its high resolution, FT-ICR can be used for the analysis of complex mixtures. FT-ICR, coupled to ESI, is also being employed in the study of protein interactions and protein conformations. A high-throughput, large scale proteomics approach involving FT-ICR has recently been developed by Smith et al.

Peptide Fragmentation

As peptide ions are introduced into the collision chamber, they interact with the collision gas (usually nitrogen or argon) and undergo fragmentation primarily along the peptide backbone. Since peptides can undergo multiple types of fragmentation, nomenclature has been created to indicate what type of ions has been generated. If, after peptide bond cleavage, the charge is maintained on the N-terminus of the ion, it is designated ab-ion, whereas if the charge is maintained on the C terminus, it is a y-ion. The difference in mass between adjacent y- or b-ions corresponds to that of an amino acid. This can be used to identify the amino acid and hence the peptide sequence, with the exception of isoleucine and leucine, which are identical in mass and therefore indistinguishable. Both y- and b-type ions can also eliminate NH_3 (17 Da), H_2O (18 Da) and CO (28 Da), resulting in pairs of signals observed in the mass spectrum. In addition to fragmentation along the peptide backbone, cleavage can occur along amino acid side chains, and this information can be used to distinguish isoleucine and leucine.

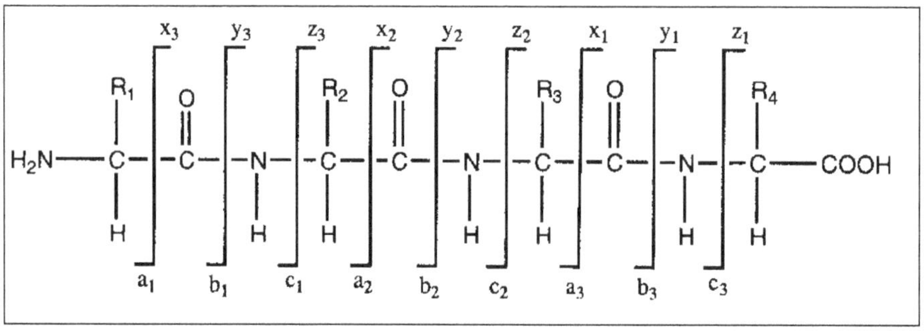

Figure shows Peptide ion fragmentation nomenclature. Low-energy collisions promote fragmentation of a peptide primarily along the peptide backbone. Peptide fragmentation which maintains the charge on the C terminus is designated a y-ion, whereas fragmentation which maintains the charge on the N terminus is designated a b-ion. Additional types of fragmentation are also indicated.

Database Utilization

Databases allow protein structural information harvested from Edman sequencing or MS to be used for protein identification. The goal of database searching is to be able to quickly and accurately identify large numbers of proteins. The success of database searching depends on the quality of the data obtained in the mass spectrometer, the quality of the database searched, and the method used to search the database. What is the best way to identify an unknown protein? What type of database search engine should be used?

Peptide Mass Fingerprinting Database Searching

One method of protein identification is peptide mass fingerprinting. In this method, the masses of peptides obtained from the proteolytic digestion of an unknown protein are compared to the predicted masses of peptides from the theoretical digestion of proteins in a database. If enough peptides from the real mass spectrum and the theoretical one overlap, protein identification can be made. The principal advantage of peptide mass fingerprinting is speed. The analysis and database search can be fully automated. The single biggest disadvantage of peptide mass fingerprinting is ambiguity in protein identification. This is because of peptide mass redundancy. For example, a peptide of 5 amino acids can have the same mass by simple rearrangement of its constitutive amino acids; e.g., peptide VAGSE has the same mass as AVGSE or AEVGS and so on. For this technique to be successful, the masses of a large number of peptides must be obtained to provide enough specificity in the search, and this is not always possible. Mass redundancy occurs with greater frequency in large genomes. Moreover, peptide mass fingerprinting is effective only in the analysis of proteins from organisms whose genome is small, completely sequenced, and well annotated. It has limited use against un-annotated or un-translated DNA databases such as the human genome. Because mass fingerprinting is not error tolerant, several factors in addition to mass redundancy contribute to its limited use, including sequencing errors, conservative substitutions, polymorphisms, and six possible translations at the DNA level.

Another factor affecting the success of peptide mass fingerprinting is mass accuracy. Because it is critical to obtain an accurate measurement of the masses of multiple peptides, factors that alter the masses of those peptides can reduce the success of the method. One such example is the post-translational modification of proteins. If the unknown protein is extensively modified, the peptides produced from that protein will not match the unmodified protein in the database. Recent improvements in the mass accuracy of mass spectrometers have increased the success rate of protein identification by this method.

Finally, peptide mass fingerprinting does not work well with protein mixtures. As a protein mixture is converted to a mixture of peptides, it increases the complexity of the peptide mass fingerprint. The process of protein identification can be hindered if even two or three proteins are present in the sample. Several search methods have emerged to accommodate peptide mixtures in the mass spectrum. One example is a program called ProFound, which enables protein identification in simple protein mixtures. However, the lack of ability to analyze protein mixtures remains a major limitation of this method. A variety of tools for database searching now exist on the World Wide Web. The ExPASy server provides a variety of tools for proteomics and programs for protein identification. Search programs used for peptide mass fingerprinting include PepSea, PeptIdent/MultiIdent, MS-Fit, MOWSE, and ProFound.

Amino Acid Sequence Database Searching

The most specific type of database searching for protein identification uses peptide amino acid sequence. If the amino acid sequence of a peptide can be identified, it can be used to search databases to find the protein from which it was derived. One method which utilizes this information is peptide mass tag searching. In this method, a partial amino acid sequence is obtained by interpretation of the MS/MS spectrum (the sequence tag) and this information is combined with the mass of the peptide and the masses of the peptide on either side of the sequence tag where the sequence is not known.

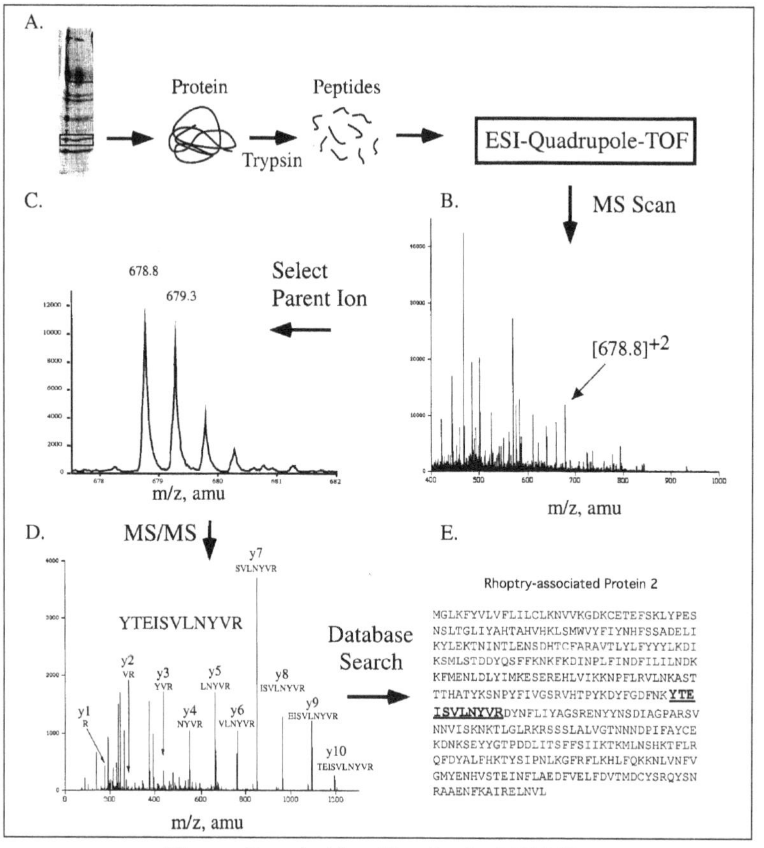

Figure: Protein identification by MS/MS.

Figure shows (A) Protein from P. falciparum was resolved on a one-dimensional polyacrylamide gel, excised, and in-gel digested with trypsin. The resulting peptides were ionized by electrospray and analyzed by a Quadrupole-TOF mass spectrometer. (B) The MS spectrum produced was scanned, and a parent ion of 678.8 was selected for fragmentation. (C) Enlargement of the parent ion peak at 678 shown in panel B. The multiplet of peaks is due to the contribution in mass from the naturally occurring isotope 13C. A mass difference between the peaks of 0.5 Da indicates that the peptide is doubly charged. (D) MS/MS scan of the 678 parent ion and analysis of the daughter ions produced. All y-ions (except for y-11) produced from fragmentation of the peptide are shown. (E) Identification of rhoptry-associated protein-2 using BioAnalyst software.

Also included in the search is the type of protease used to produce the peptides. Peptide mass tag searching is a more specific tool for protein identification than peptide mass fingerprinting. In addition, one of the biggest advantages of utilizing MS/MS to obtain peptide amino acid sequence is that, unlike peptide mass fingerprinting, it is compatible with protein mixtures. The ability to identify proteins in mixtures is one of the great advantages of using MS as a protein identification tool.

For example, in our laboratory we frequently identify multiple proteins from what appears to be a single band on an SDS-gel. In fact, in the majority of proteomics experiments, proteins are present in mixtures at the time of analysis. The major disadvantage of performing MS/MS is that the process is not easily automated. As a result, considerable time is expended in performing the analysis and interpreting the mass spectrum. Although computer programs can assist in the interpretation of the spectrum, they currently are not able to make accurate assignments without some guidance. In addition, when searching a database with peptide mass tags, there is a lack of flexibility in the search programs. If a single mistake is made in the assignment of a y- or b-ion (which can happen quite frequently), the amino acid sequence will be incorrect and the database search will bring up irrelevant proteins. Often it is necessary to confirm that the peptide sequence obtained from the database matches the sequence obtained in the mass spectrometer. This can be done by performing a theoretical fragmentation of the peptide from the database and comparing the two mass spectra. Additional clues can also be used, such as verifying if the peptide obtained from the database ends in amino acids consistent with the type of protease used.

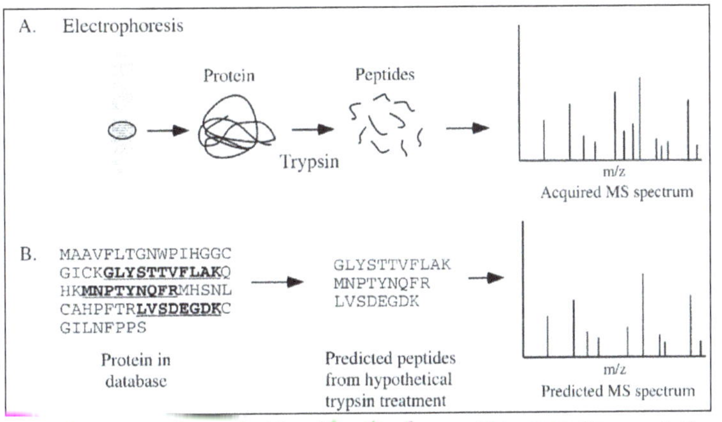

Figure: Strategy of protein identification by peptide mass fingerprinting.

Figure shows (A) The unknown protein is excised from a gel and converted to peptides by the action of a specific protease. The mass of the peptides produced is then measured in a mass spectrometer. (B) The mass spectrum of the unknown protein is searched against theoretical mass spectra produced by computer generated cleavage of proteins in the database.

De Novo Peptide Sequence Information

Another approach to protein identification is to obtain de novo sequence data from peptides by MS/MS and then use all the peptide sequences to search appropriate databases. Multiple peptide sequences can be used for protein identification by searching databases with the FASTS program (Mackey et al., submitted). The single biggest advantage of this method is the capability of searching peptide sequence information across both DNA and protein databases. This is because the search engine utilized exhibits a certain amount of flexibility in the assignment of protein scores. This search method is useful for organisms that do not have well-annotated databases such as Xenopus laevis or human. However, because this method requires several peptide amino acid sequences of 3 or 4 amino acids, it is not the first choice for peptide identification. Rather, the much faster methods of peptide mass fingerprinting or peptide mass tag searching can be used first. If these search methods fail, de novo sequence information can be obtained and used to identify the protein.

Uninterpreted MS/MS Data Searching

A large number of programs are now available for the identification of proteins by using uninterpreted MS/MS data. Examples include programs such as Mascot, SONAR, and SEQUEST.

However, searches against un-annotated or un-translated DNA databases with uninterpreted MS/MS data are likely to suffer from the same pitfalls associated with mass fingerprinting. In particular, polymorphisms, sequencing errors, and conservative substitutions will probably contribute to failure to accurately identify a protein. The development of uninterpreted MS/MS search algorithms that are error tolerant may overcome some of these shortcomings, provided that they assign some form of statistical scoring to the identified proteins.

Proteomics Applications

The single most common application of proteomics is protein identification. Most investigators use proteomics approaches to isolate and display proteins based on their own specific criteria and then identify the proteins. Protein identification provides immediate information that will direct subsequent experimentation. For example, the identity of a protein can reveal an expected result, validate a proteomics approach, provide completely unexpected information, or reveal that your biochemical method is not working at all. In recent years, as the technology of MS has improved, there has been a de-emphasis on the "front-end" of proteomics experiments compared to data analysis. This can result in the isolation of hundreds of irrelevant proteins for identification, consuming both time and effort. Our general strategy is to devise techniques that enrich for low-abundance proteins and then analyze only the proteins that appear on differential display or are isolated by affinity chromatography. To accomplish this, we use affinity columns and other strategies to select for protein targets. In each case, protein samples are subjected to a series of pre-columns and high-stringency washes to remove nonspecific proteins. This reduces the number of irrelevant proteins for analysis.

Characterization of Protein Complexes

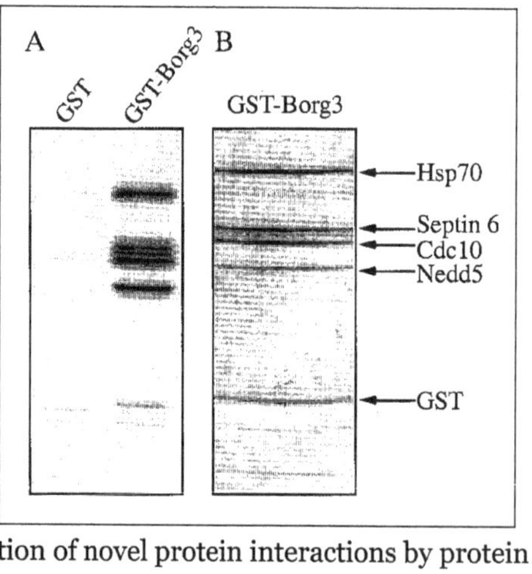

Figure: Identification of novel protein interactions by protein coprecipitation.

Figure shows: (A) Pull-down experiment with a control (GST) or target (GST-Borg3) protein using 35S-labeled NIH 3T3 cell lysate. (B) Largescale affinity purification of GST-Borg3 from the NIH 3T3 lysate. Individual proteins were microsequenced by mixed-peptide sequencing and identified by database searching with the FASTF algorithm.

Many laboratories are now engaged in an effort to characterize protein complexes by MS. Examples include Link et al. utilizing multidimensional LC and MS/MS to identify proteins or Mann and colleagues identifying proteins present after immunoprecipitation of protein complexes. Recently, Macara, Haystead, and coworkers used MS to identify interacting proteins with the Cdc42 effector, Borg3. In this case, the "bait" protein, Borg3, was produced as a glutathione S-transferase (GST) fusion in E. coli and then mixed with NIH 3T3 cell lysate. Four interacting proteins were identified by mixed-peptide sequencing: heat shock protein Hsp70 and three septins including Septin6, Cdc10, and Nedd5. None of these proteins were present in the GST-only control sample. Although the interaction with Hsp70 was not pursued, it was shown from coimmunoprecipitation studies that endogenous Borg3 interacts with endogenous Cdc10 and Nedd5. Additional proof from expression and structure-function studies confirmed a role for the Borg proteins as regulators of septin organization. It should be noted that although several proteins were quickly identified as Borg3 interactors by the pull-down experiment, it took several more months of work to confirm this interaction.

Protein Expression Profiling

The largest application of proteomics continues to be protein expression profiling. Through the use of two-dimensional gels or novel techniques such as ICAT, the expression levels of proteins or changes in their level of modification between two different samples can be compared and the proteins can be identified. This approach can facilitate the dissection of signaling mechanisms or identify disease-specific proteins.

Expression Profiling by Two-Dimensional Electrophoresis

Currently, the majority of protein expression profiling studies are performed by 2-DE. Several diseases have been studied, including heart disease and cancer. Cancer cells are good candidates for proteomics studies because they can be compared to their non-transformed counterparts. Analysis of differentially expressed proteins in normal versus cancer cells can (i) identify novel tumor cell biomarkers that can be used for diagnosis, (ii) provide clues to mechanisms of cancer development, and (iii) identify novel targets for therapeutic intervention. Protein expression profiling has been used in the study of breast, esophageal, bladder and prostate cancer. From these studies, tumor-specific proteins were identified and 2-D protein expression databases were generated. Many of these 2-D protein databases are now available on the World Wide Web.

Isotope-Coded Affinity Tags

Recently, a novel method for protein expression profiling was introduced that does not depend on the separation of proteins by 2-DE. This method is known as isotope-coded affinity tags (ICAT) and relies on the labeling of protein samples from two different sources with two chemically identical reagents that differ only in mass as a result of isotope composition. Differential labeling of samples by mass allows the relative amount of protein between two samples to be quantitated in the mass spectrometer. Cell extract from two different samples is reacted with one of two forms of the ICAT reagent, an isotopically light form in which the linker contains eight hydrogens or a heavy form in which the linker contains eight deuterium atoms. The ICAT reagent reacts with cysteine residues in proteins via a thiol-reactive group and contains a biotin moiety to facilitate purification. Peptides are recovered on the basis of the biotin tag by avidin affinity chromatography and are then analyzed by MS.

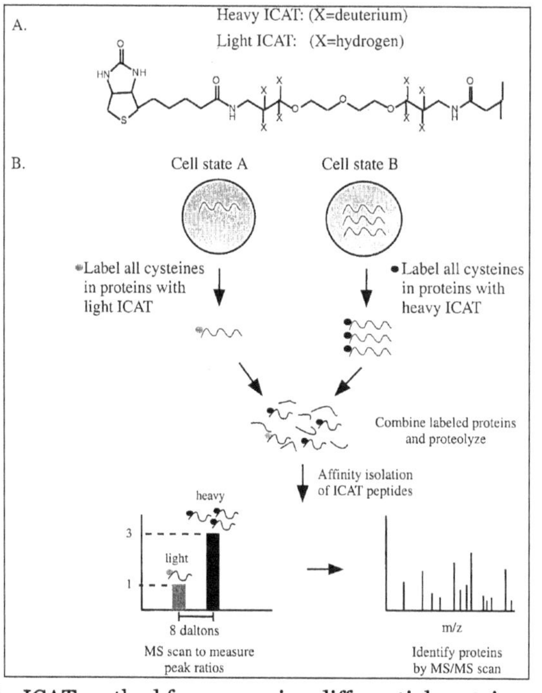

Figure: The ICAT method for measuring differential protein expression.

Figure shows: (A) Structure of the ICAT reagent. ICAT consists of a biotin affinity group, a linker region that can incorporate heavy (deuterium) or light (hydrogen) atoms, and a thiol-reactive end group for linkage to cysteines. (B) ICAT strategy. Proteins are harvested from two different cell states and labeled on cysteine residues with either the light or heavy form of the ICAT reagent. Following labeling, the two protein samples are mixed and digested with a protease such as trypsin. Peptides labeled with the ICAT reagent can be purified by virtue of the biotin tag by using avidin chromatography. Following purification, ICAT-labeled peptides can be analyzed by MS to quantitate the peak ratios and proteins can be identified by sequencing the peptides with MS/MS.

The difference in peak heights between heavy and light peptide ions directly correlates with the difference in protein abundance in the cells. Thus, if a protein is present at a threefold higher level in one sample, this will be reflected in a threefold difference in peak heights. Following quantitation of the peptides, they can be fragmented by MS/MS and the amino acid sequence can be obtained. Thus, using this approach, proteins can be identified and their expression levels can be compared in the same analysis.

The single biggest advantage of this method is the elimination of the 2-D gel for protein quantitation. As a result, an increased amount of sample can be used to enrich for low abundance proteins. Alternatively, the cell lysate can be fractionated prior to reaction with the ICAT reagent. This can allow the enrichment of low-abundance proteins before the analysis begins. The main disadvantages are that currently this method works only for proteins containing cysteine, even though this includes the majority of proteins. In addition, peptides must contain appropriately spaced protease cleavage sites flanking the cysteine residues. Finally, the ICAT label is large (500 kDa) and remains with each peptide throughout the analysis. This can make database searching more difficult, especially for small peptides with limited sequence. Sensitivity may also be of concern since tagged peptides derived from low-copy proteins are likely to be poorly recovered during the affinity step as a result of nonspecific interactions with avidin-Sepharose. Studies have been performed to optimize the labeling of proteins with the ICAT reagent.

Protein Arrays

Protein arrays are undergoing rapid development for the detection of protein-protein interactions and protein expression profiling. Recently, protein microarrays were created using ordinary laboratory equipment. Proteins were immobilized by being covalently attached to glass microscope slides, and the protein microarrays were shown to be capable of interacting with other proteins, small molecules, and enzyme substrates. In another report, 5,800 yeast proteins were expressed and printed onto microscope slides. These protein microarrays were used to identify novel calmodulin- and phospholipid-interacting proteins. These reports indicate that protein arrays hold great promise for the global analysis of protein-protein and protein-ligand interactions. Undoubtedly, these arrays will improve as the technology for their creation is developed and refined.

Proteomics Approach to Protein Phosphorylation

Posttranslational modification of proteins is a fundamental regulatory mechanism, and characterization of protein modifications is paramount for understanding protein function. MS is one of the most powerful tools for the analysis of protein modifications because virtually any type of protein modification can be identified. Although we focus here on protein phosphorylation, the analysis of other types of protein modification by MS has been described. Protein phosphorylation is one of the most common of all protein modifications and has been found in nearly all cellular processes. MS can be used to identify novel phosphoproteins, measure changes in the phosphorylation state of proteins in response to an effector, and determine phosphorylation sites in proteins. Identification of phosphorylation sites can provide information about the mechanism of enzyme regulation and the protein kinases and phosphatases involved. A proteomics approach to protein phosphorylation has the advantage that instead of studying changes in the phosphorylation of a single protein in response to some perturbation, one can study all the phosphoproteins in a cell (the phosphoproteome) at the same time. A common approach to studying protein phosphorylation events is the use of in vivo labeling of phosphoproteins with inorganic 32P. The phosphoproteomes of cells that differ in some way (e.g., normal versus diseased) can be analyzed by growing cells in inorganic 32P and creating cell lysates. Changes in the phosphorylation state of proteins can then be examined by 2-DE and autoradiography. Proteins of interest are excised from the gel and micro-sequenced by MS. A major limitation of this approach is that while many phosphorylated proteins can be visualized by autoradiography, they cannot be identified because of their low abundance. One solution to this problem is enrichment of the phosphoproteome.

Phosphoprotein Enrichment

Enrichment of the phosphoproteome of a cell can allow the identification of low-copy phosphoproteins that would otherwise go undetected. In one approach, phosphoproteins were enriched by conversion of phosphoserine residues to biotinylated residues. This method is an extension of techniques originally developed by Hielmeyer and colleagues and more recently by our laboratory for the identification of phosphorylation sites using Edman sequencing. Following derivatization, proteins that were formerly phosphorylated can be isolated by avidin affinity chromatography. Proteins immobilized on avidin beads can then be eluted with biotin, theoretically resulting in the isolation of the entire phosphoserine proteome. By increasing the amount of cell lysate used for avidin affinity chromatography, low-abundance phosphoproteins can be enriched. However, this technique does not work for phosphotyrosine and the reactivity of phosphothreonine by this method is very poor. Tyrosine-phosphorylated proteins can be isolated by the use of antiphosphotyrosine antibodies. As an alternative, another method for phosphopeptide enrichment was devised to allow the recovery of proteins phosphorylated on serine, threonine, and tyrosine. In this method, a protein or mixture of proteins is digested to peptides with a protease and then subjected to a multistep procedure for the conversion of phosphoamino acids into free sulfhydryl groups. To capture the derivatized peptides, the free sulfhydryl groups in the peptides are then reacted with iodoacetyl groups immobilized on glass beads. Using this method, several phosphopeptides were recovered from -casein and from a yeast cell extract, although it was unclear whether all the proteins isolated from the yeast extract were phosphoproteins.

Enrichment of the phosphoproteome can also be combined with protein profiling by 1- or 2-DE. In this way, changes in protein amount observed on electrophoresis will reflect the level of protein phosphorylation. Recently, the principle of protein quantitation by ICAT has been combined with phosphoprotein enrichment. This was accomplished by the introduction of isotopic label into ethanedithiol, the reagent used to convert the alkene created by -elimination of phosphoserine into a free sulfhydryl group. In this way, the differences in the amount of phosphoproteins in extracts can be analyzed quantitatively in the mass spectrometer. It should be noted that because of the chemistry used in both of these methods, these techniques are relatively insensitive and require tens of picomoles of phosphoprotein.

Phosphorylation Site Determination by Edman Degradation

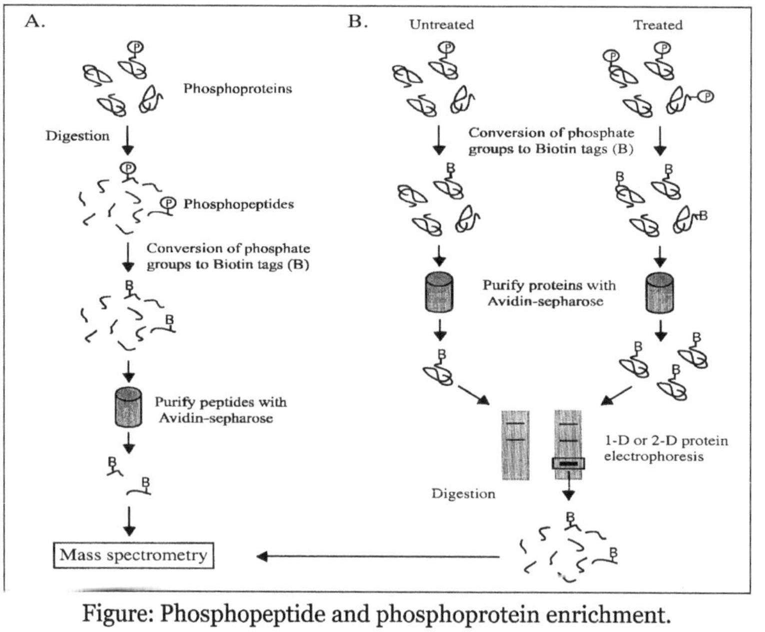

Figure: Phosphopeptide and phosphoprotein enrichment.

Figure shows: (A) Enrichment of phosphopeptides. Phosphoproteins are digested with a protease, and the phosphate groups are converted to biotin tags. Once biotinylated, the peptides can be selectively recovered with avidin-Sepharose and analyzed by MS. (B) Differential display of phosphoproteins. The phosphate groups present in proteins derived from two different samples are converted to biotin tags, and the phosphoproteins are purified on avidin-Sepharose in an identical manner. The phosphoproteins are then compared by 1- or 2-DE, and the target proteins are digested and analyzed by MS. Edman sequencing is still a widely used method for determining phosphorylation sites in proteins labeled with 32P, either in vitro or in vivo. This is because sites can be determined at the sub-femtomolar level if enough radioactivity can be incorporated into the phosphoprotein of intrest, this can be as little as 1,000 cpm (not ideal). Briefly, a 32P-labeled protein is digested with a protease and the resulting phosphopeptides are separated and purified by reversephase HPLC or thin-layer chromatography (TLC). The isolated peptides are then cross-linked via their C terminito an inert membrane (e.g. Immobilon P; PerSeptive Biosystems). The radioactive membrane is subjected to several rounds of Edman cycles, and radioactivity is collected after the cleavage step. The released 32P is counted in a scintillation counter. This method positionally places the phosphoamino acid within the sequenced phosphopeptide. Of course, this is meaningful only if the sequence of the phosphopeptide is already known. In addition, the analysis ceases to become quantitative beyond 30 Edman cycles (even with efficient, modern Edman machines) due to well-understood issues with repetitive yield associated with Edman chemistry.

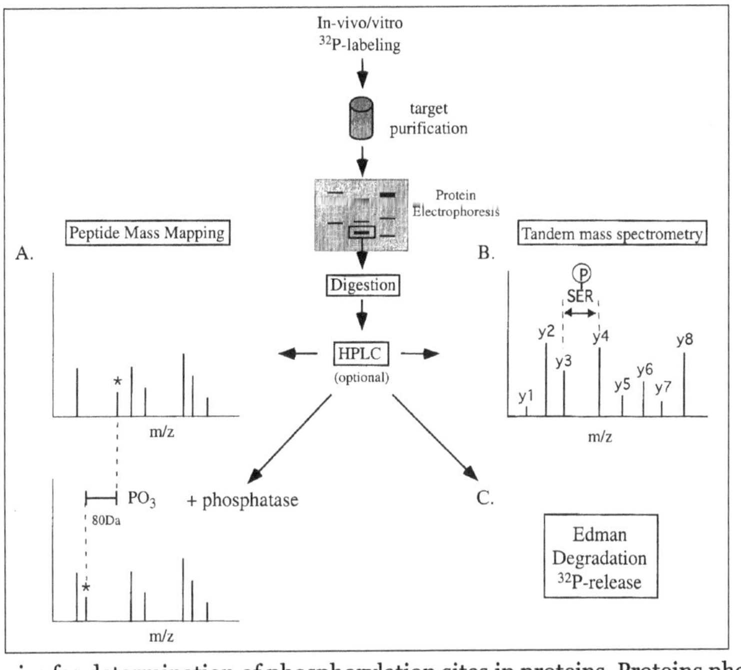

Figure: Strategies for determination of phosphorylation sites in proteins. Proteins phosphorylated in vitro or in vivo can be isolated by protein electrophoresis and analyzed by MS.

Figure shows: (A) Identification of phosphopeptides by peptide mass fingerprinting. In this method, phosphopeptides are identified by comparing the mass spectrum of an untreated sample to that of a sample treated with phosphatase. In the phosphatase-treated sample, potential phosphopeptides are identified by a decrease in mass due to loss of a phosphate group (80 Da). (B) Phosphorylation sites can be identified by peptide sequencing using MS/MS. (C) Edman degradation can be used to monitor the release of inorganic 32P to provide information about phosphorylation sites in peptides.

In CRP analysis, one requires only that the sequence of the protein be known. Purification and sequencing of individual peptides is not required. Radiolabeled proteins (isolated following immunoprecipitation from 32P-labeled cells, for example) are cleaved at predetermined residues by the action of a protease. The phosphopeptides are then separated by HPLC or TLC (if only one site is present, no peptide separation is required), cross-linked to the inert membrane, and carried through 25 to 30 Edman cycles. The sequence of the target protein is entered into the CRP program. This program predicts how many Edman cycles are required to cover 100% of all the serines, threonines, and tyrosines from the site of cleavage. Generally, one round of CRP analysis narrows the number of possible sites to 5 to 10 for most proteins. Phosphoamino acid analysis can be used to reduce the number of possibilities still further. The CRP analysis is then repeated following cleavage with a second protease (usually one cutting at R, but M and F are alternatives). The second round of CRP usually unambiguously localizes the phosphoamino acid to one possible site. The technique does not work if sites are more than 30 amino acids away from all possible cleavage sites. The finding that CRP analysis is not applicable may in itself confine a phosphorylation site to a segment of the protein that is likely to produce very large proteolytic fragments.

Phosphorylation Site Determination by Mass Spectrometry

Because of its sensitivity, MS can allow the direct sequencing of phosphopeptides, resulting in unambiguous phosphorylation site identification. Identification of phosphorylation sites in proteins provides several unique challenges for the mass spectrometrist. For example, unlike in protein identification, where analysis of any peptide within the protein can be informative, phosphorylation site analysis requires that the phosphorylated peptide be analyzed. This means that considerably more protein is required for analysis. In addition, phosphorylation can alter the cleavage pattern of a protein and the resulting phosphopeptides may require different purification methods. To isolate and purify the phosphopeptides of interest, it may be necessary to alter the way in which the phosphoprotein is digested and to alter the pH or the chromatographic material used for peptide purification.

(i) Phosphopeptide sequencing by MS/MS: A combination of HPLC, Edman degradation, and phosphopeptide sequencing by MS/MS provides the best results for phosphorylation site determination. Following excision and digestion of a 32P-labeled protein, the peptides are resolved by HPLC. By monitoring HPLC fractions for radioactivity, the phosphopeptides can be selected for analysis. This reduces the complexity of the peptide mixture before MS is performed and facilitates phosphopeptide identification. Phosphopeptides can be identified from a mixture of peptides by a method known as precursor ion scanning. In this method, the second mass analyzer in the mass spectrometer is set at the mass of the reporter ion for the phospho group (PO_3^-) of m/z =79. Peptides are sprayed under neutral or basic conditions, and phosphopeptides are identified in the precursor ion scan only if their fragmentation yields an ion of m/z = 79. Once a phosphopeptide is identified, the peptide mixture is sprayed under acidic conditions and the phosphopeptide is sequenced by conventional tandem MS/MS. On fragmentation of the phosphopeptide, phosphoserine can be identified by the formation of dehydroalanine (69 Da), the β-elimination product of phosphoserine. Similarly, phosphothreonine can be identified by the formation of its β-elimination product, dehydroamino-2-butyric acid at 83 Da.

(ii) Analysis of phosphopeptides by MALDI-TOF: MALDITOF mass spectrometry can also be used to identify phosphopeptides. When phosphorylated peptides are subjected to ionization by MALDI, phosphate groups are frequently liberated from the peptides. This is the case for phosphoserine- and phosphothreonine-containing peptides, which can liberate HPO_3 or H_3PO_4, resulting in a neutral loss of 80 and 98 Da, respectively. Careful examination of the TOF spectrum for differences in peptide masses of 80 Da that are not found in the unphosphorylated peptide control can identify phosphopeptides. Phosphopeptides can also be identified by treating one of two identical samples with protein phosphatase to liberate phosphate groups. Once a phosphopeptide is identified, it can be sequenced by MS/MS for identification of the phosphorylation site.

Yeast Genomics and Proteomics

One of the most exciting applications of proteomics involves combining this technology with the power of yeast genetics to delineate signaling events in vivo. In one study, we identified physiological substrates for the Glc7p-Reg1p complex by examining the effects of deletion of the REG1 gene on the yeast phosphoproteome. In S. cerevisiae, PP-1 (Glc7p) and its binding protein, Reg1p, are essential for the regulation of glucose repression pathways. The target for this phosphatase complex was not known. Analysis by 2-D phosphoprotein mapping identified two distinct proteins that were greatly increased in phosphate content in reg1 Δ mutants. Mixed-peptide sequencing identified these proteins as hexokinase II (Hxk2p) and the E1α subunit of pyruvate dehydrogenase. We then went on to validate these findings in a comprehensive biochemical study. Consistent with increased phosphorylation of Hxk2p in response to REG1 deletion, fractionation of yeast extracts by anion-exchange chromatography identified an Hxk2p phosphatase activity in wild-type strains that was selectively lost in the reg1 Δ mutant. Having carried out these studies, we attempted to rescue the reg1 Δ phosphoprotein phenotype by overexpressing both wild-type and mutant Reg1p in the deletion strains. Here, both the phosphorylation state of Hxk2p and Hxk2p phosphatase activity were restored to wild-type levels in the reg1 Δ mutant by expression of a LexA-Reg1p fusion protein. In contrast, expression of a LexA-Reg1p protein containing mutations at phenylalanine in a putative PP-1C (the catalytic subunit) binding site motif (K/R)(X)(I/V)XF was unable to rescue Hxk2p dephosphorylation in intact yeast or restore Hxk2p phosphatase activity. These results demonstrate that Reg1p targets PP-1C to dephosphorylate Hxk2p in vivo and that the peptide motif (K/R)(X)(I/V)XF is necessary for its PP-1 targeting function. These studies therefore demonstrate how a proteomics approach can be used to first identify enzyme targets in cells and then direct all further analysis to verify the findings. It should be pointed out that often 6 to 12 months of work ensues following the initial sequencing of the targeted proteins. Nevertheless, clearly a combined proteomics and genetics approach greatly enhances one's ability to directly answer key biological questions. We believe that a similar strategy could be adopted with transgenic or knockout mouse work, particularly in cases where there is no obvious phenotype.

Proteome Mining

Proteome mining is a functional proteomics approach used to extract protein information from the analysis of specific subproteomes. The principles of proteome mining are based on the assumption that all drug-like molecules selectively compete with a natural cellular ligand for a binding site on a protein target. In a proteome mine, natural ligands are immobilized on beads at high density and in an orientation that sterically favors interaction with their protein targets. The immobilized ligand is then exposed to whole-animal or tissue extract, and bound proteins are evaluated for specificity by protein sequencing. Micro-sequencing of the proteins that were eluted with free ATP demonstrated that the nucleotide selectively recovered purine binding proteins including protein kinases, dehydrogenases, various purine-dependent metabolic enzymes, DNA ligases, heat shock proteins, and a variety of miscellaneous ATP-utilizing enzymes. This immobilized proteome represents 4% of the expressed eukaryotic genome.

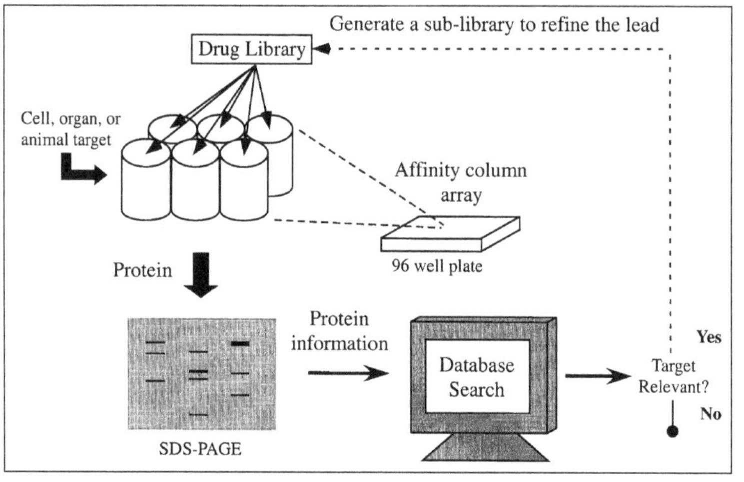

Figure shows Proteome-mining strategy, Proteins are isolated on affinity column arrays from a cell line, organ, or animal source and purified to remove nonspecific adherents. Then, compound libraries are passed over the array and the proteins eluted are analyzed by protein electrophoresis. Protein information obtained by MS or Edman degradation is then used to search DNA and protein databases. If a relevant target is identified, a sublibrary of compounds can be evaluated to refine the lead. From this analysis, both a protein target and a drug lead can be simultaneously identified.

We have utilized this captured proteome (the purine binding cassette proteome) to test the selectivity of purine analogs that inhibit protein kinases and stress-induced ATPases in vitro. Using a proteome-mining ATP affinity array apparatus constructed in our laboratory, sufficient biomass was applied to ensure the recovery, per column, of 1 fmol of any protein expressed at 100 copies/cell (107 cells). After washing, each column in the array is eluted in parallel with molecules from a purine-based iterative library and fractions are collected. Eluates are screened for protein, and positive fractions generally contain a single protein, a small number of structurally related proteins, or a complex mixture. Only the first two categories are sequenced, since the third resulted from elution with a nonselective inhibitor. Once one has identified an eluted protein, one has all the necessary information on how to proceed. The first decision is biological relevance. Does the eluted protein(s) in any given fraction have relevance to any human disease? If the protein has no obvious use as a drug target, it is ignored. If the protein is deemed relevant, one immediately has a lead molecule and a defined target. In cases where a single protein is eluted, the lead is likely to be selective because it had an equal opportunity to interact with the rest of the captured proteome (4% of the genome).

Selectivity can be tested by increasing the concentration of the lead compound during elution from nano-molar to micro-molar. Information concerning potential toxicity can be gained by sequencing other proteins that are simultaneously eluted or eluted at higher concentrations. If some of these are undesirable targets, iterative substitutions can be made around the lead scaffold to improve selectivity. Proof of principle of this technology was obtained by using an iterative library derived from the heat shock protein 90 inhibitor geldanamycin, and a new physiological target, ADE2, was identified. Screening Combichem libraries through a proteome-mining approach exploits the serendipitous nature of drug discovery to its maximum, merely because it accelerates the hit rate over a conventional screen by a factorial of the proteome that is bound. In the case of purine binding proteins, this may be several hundredfold. Protein micro-sequencing, the data contained within the various genome projects, and the ability to instantly search the literature for relevance enable one to interpret the outcomes in a rationale way.

We are currently using proteome mining to discover new antimalarial drugs that target purine binding proteins in the blood stage of infection. The P. falciparum (blood stage) and human red blood cell purine binding proteome are captured on ATP affinity arrays and simultaneously screened against purine-based combinatorial libraries. Combining both proteomes enables the selectivity and potential toxicity of a lead molecule to be measured early in the discovery process. Micro-sequencing enables human proteins to be readily discriminated from malarial ones. An additional benefit of mining the entire malarial purine binding cassette proteome is that multiple leads and their targets will be identified. Combined therapies that target multiple genes simultaneously are likely to exert such tremendous selective pressure on the targeted pathogen that it cannot develop resistance.

Challenges for Proteomics

The study of proteins, in contrast to that of DNA, presents a number of unique challenges. For example, there is no equivalent of PCR for proteins, so the analysis of low-abundance proteins remains a major challenge. In addition, in protein interaction studies, native conformations of proteins must be maintained to obtain meaningful results. Can proteins be studied on a large scale with speed, sensitivity, and reliability? In the last several years, recognition of the limitations of proteomics is beginning to point the field in new directions.

Although the technology for the analysis of proteins is rapidly progressing, it is still not feasible to study proteins on a scale equivalent to that of the nucleic acids. Most of proteomics relies on methods, such as protein purification or PAGE, that are not high-throughput methods. Even performing MS can require considerable time in either data acquisition or analysis. Although hundreds of proteins can be analyzed quickly and in an automated fashion by a MALDI-TOF mass spectrometer, the quality of data is sacrificed and many proteins cannot be identified. Much higher quality data can be obtained for protein identification by MS/MS, but this method requires considerable time in data interpretation. In addition, to access un-annotated DNA databases across species, these algorithms should be error tolerant to allow for sequencing errors, polymorphisms, and conservative substitutions. New technologies will have to emerge before protein analysis on a large-scale (such as mapping the human proteome) becomes a reality.

Another major challenge for proteomics is the study of low abundance proteins. In some eukaryotic cells, the amounts of the most abundant proteins can be 106 -fold greater than those of the low-abundance proteins. Many important classes of proteins (that may be important drug targets) such as transcription factors, protein kinases, and regulatory proteins are low-copy proteins. These low-copy proteins will not be observed in the analysis of crude cell lysates without some purification. Therefore, new methods must be devised for subproteome isolation. Despite these limitations, proteomics, when combined with other complementary technologies such as molecular biology, has enormous potential to provide new insight into biology. The ability to study complex biological systems in their entirety will ultimately provide answers that cannot be obtained from the study of individual proteins or groups of proteins.

Protein

Protein is the highly complex substance that is present in all living organisms. Proteins are of great nutritional value and are directly involved in the chemical processes essential for life. The importance of proteins was recognized by chemists in the early 19th century, including Swedish chemist Jöns Jacob Berzelius, who in 1838 coined the term protein, a word derived from the Greek prōteios, meaning "holding first place." Proteins are species-specific; that is, the proteins of one species differ from those of another species. They are also organ-specific; for instance, within a single organism, muscle proteins differ from those of the brain and liver.

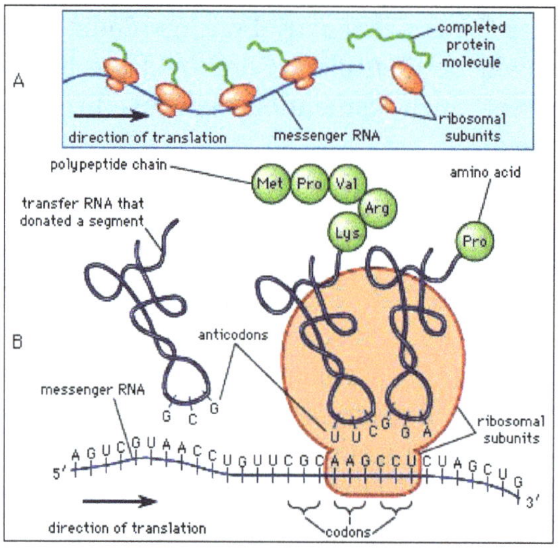

Figure: Protein synthesis.

A protein molecule is very large compared with molecules of sugar or salt and consists of many amino acids joined together to form long chains, much as beads are arranged on a string. There are about 20 different amino acids that occur naturally in proteins. Proteins of similar function have similar amino acid composition and sequence. Although it is not yet possible to explain all of the functions of a protein from its amino acid sequence, established correlations between structure and function can be attributed to the properties of the amino acids that compose proteins.

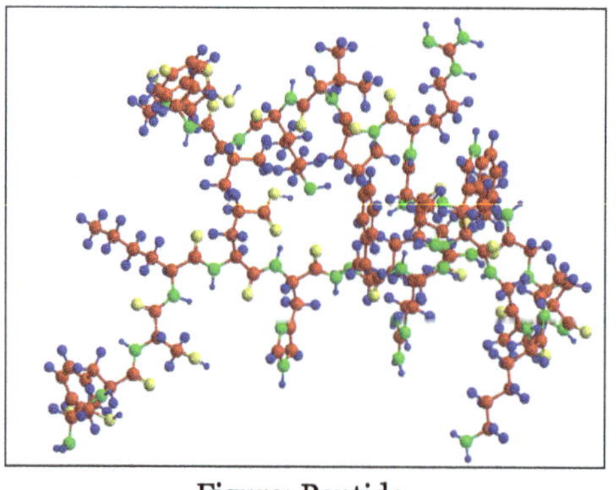

Figure: Peptide.

Plants can synthesize all of the amino acids; animals cannot, even though all of them are essential for life. Plants can grow in a medium containing inorganic nutrients that provide nitrogen, potassium, and other substances essential for growth. They utilize the carbon dioxide in the air during the process of photosynthesis to form organic compounds such as carbohydrates. Animals, however, must obtain organic nutrients from outside sources. Because the protein content of most plants is low, very large amounts of plant material are required by animals, such as ruminants (e.g., cows), that eat only plant material to meet their amino acid requirements. Nonruminant animals, including humans, obtain proteins principally from animals and their products—e.g., meat, milk, and eggs. The seeds of legumes are increasingly being used to prepare inexpensive protein-rich food.

Figure: Legume; amino acid. Legumes—such as beans, lentils, and peas—are high in protein and contain many essential amino acids.

The protein content of animal organs is usually much higher than that of the blood plasma. Muscles, for example, contain about 30 percent protein, the liver 20 to 30 percent, and red blood cells 30 percent. Higher percentages of protein are found in hair, bones, and other organs and tissues with low water content. The quantity of free amino acids and peptides in animals is much smaller than the amount of protein; protein molecules are produced in cells by the stepwise alignment of amino acids and are released into the body fluids only after synthesis is complete.

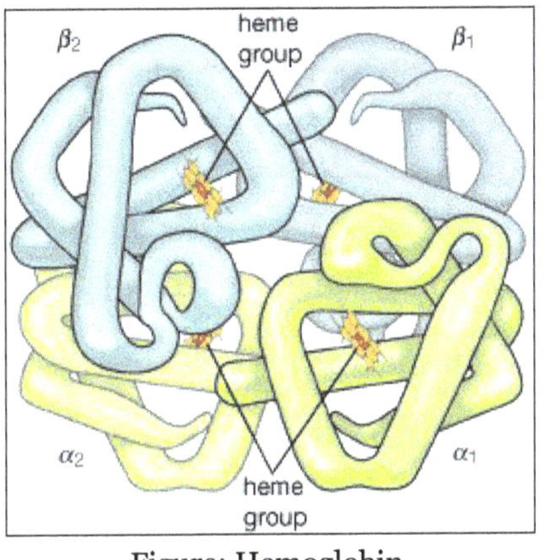

Figure: Hemoglobin.

The high protein content of some organs does not mean that the importance of proteins is related to their amount in an organism or tissue; on the contrary, some of the most important proteins, such as enzymes and hormones, occur in extremely small amounts. The importance of proteins is related principally to their function. All enzymes identified thus far are proteins. Enzymes, which are the catalysts of all metabolic reactions, enable an organism to build up the chemical substances necessary for life—proteins, nucleic acids, carbohydrates, and lipids—to convert them into other substances, and to degrade them. Life without enzymes is not possible. There are several protein hormones with important regulatory functions. In all vertebrates, the respiratory protein hemoglobin acts as oxygen carrier in the blood, transporting oxygen from the lung to body organs and tissues. A large group of structural proteins maintains and protects the structure of the animal body.

Hemoglobin is a protein made up of four polypeptide chains (α1, α2, β1, and β2). Each chain is attached to a heme group composed of porphyrin (an organic ringlike compound) attached to an iron atom. These iron-porphyrin complexes coordinate oxygen molecules reversibly, ability directly related to the role of hemoglobin in oxygen transport in the blood.

General Structure and Properties of Proteins

The Amino Acid Composition of Proteins

The common property of all proteins is that they consist of long chains of α-amino (alpha amino) acids. The general structure of α-amino acids is shown in . The α-amino acids are so called because the α-carbon atom in the molecule carries an amino group ($-NH_2$); the α-carbon atom also carries a carboxyl group ($-COOH$).

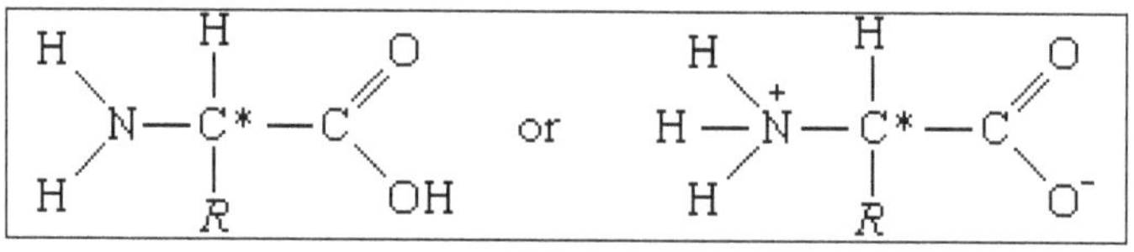

In acidic solutions, when the pH is less than 4, the $-COO$ groups combine with hydrogen ions (H+) and are thus converted into the uncharged form ($-COOH$). In alkaline solutions, at pH above 9, the ammonium groups ($-NH+3$) lose a hydrogen ion and are converted into amino groups ($-NH_2$). In the pH range between 4 and 8, amino acids carry both a positive and a negative charge and therefore do not migrate in an electrical field. Such structures have been designated as dipolar ions, or zwitterions (i.e., hybrid ions). Although more than 100 amino acids occur in nature, particularly in plants, only 20 types are commonly found in most proteins. In protein molecules the α-amino acids are linked to each other by peptide bonds between the amino group of one amino acid and the carboxyl group of its neighbour.

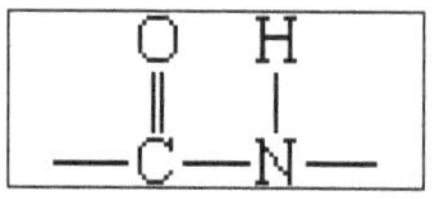

The condensation (joining) of three amino acids yields the tripeptide.

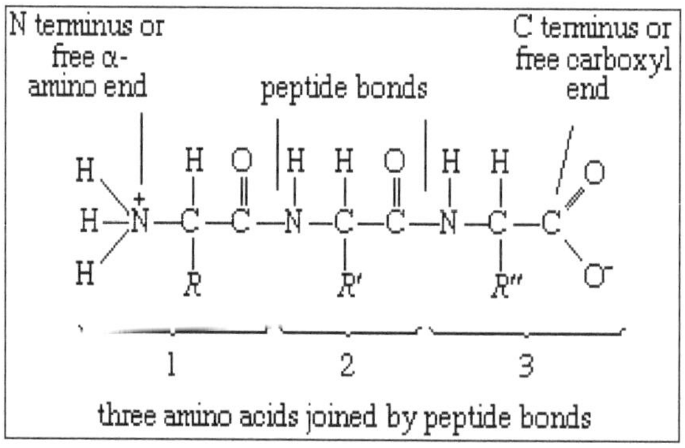

It is customary to write the structure of peptides in such a way that the free α-amino group (also called the N terminus of the peptide) is at the left side and the free carboxyl group (the C terminus) at the right side. Proteins are macromolecular polypeptides—i.e., very large molecules (macromolecules) composed of many peptide-bonded amino acids. Most of the common ones contain more than 100 amino acids linked to each other in a long peptide chain. The average molecular weight (based on the weight of a hydrogen atom as 1) of each amino acid is approximately 100 to 125; thus, the molecular weights of proteins are usually in the range of 10,000 to 100,000 daltons (one dalton is the weight of one hydrogen atom). The species-specificity and organ-specificity of proteins result from differences in the number and sequences of amino acids. Twenty different amino acids in a chain 100 amino acids long can be arranged in far more than 10100 ways (10100 is the number one followed by 100 zeroes).

Structures of Common Amino Acids

The amino acids present in proteins differ from each other in the structure of their side (R) chains. The simplest amino acid is glycine, in which R is a hydrogen atom. In a number of amino acids, R represents straight or branched carbon chains. One of these amino acids is alanine, in which R is the methyl group ($-CH_3$). Valine, leucine, and isoleucine, with longer R groups, complete the alkyl side-chain series. The alkyl side chains (R groups) of these amino acids are nonpolar; this means that they have no affinity for water but some affinity for each other. Although plants can form all of the alkyl amino acids, animals can synthesize only alanine and glycine; thus valine, leucine, and isoleucine must be supplied in the diet.

Two amino acids, each containing three carbon atoms, are derived from alanine; they are serine and cysteine. Serine contains an alcohol group ($-CH_2OH$) instead of the methyl group of alanine, and cysteine contains a mercapto group ($-CH_2SH$). Animals can synthesize serine but not cysteine or cystine. Cysteine occurs in proteins predominantly in its oxidized form (oxidation in this sense meaning the removal of hydrogen atoms), called cystine. Cystine consists of two cysteine molecules linked by the disulfide bond ($-S-S-$) that results when a hydrogen atom is removed from the mercapto group of each of the cysteines. Disulfide bonds are important in protein structure because they allow the linkage of two different parts of a protein molecule to—and thus the formation of loops in—the otherwise straight chains. Some proteins contain small amounts of cysteine with free sulfhydryl ($-SH$) groups.

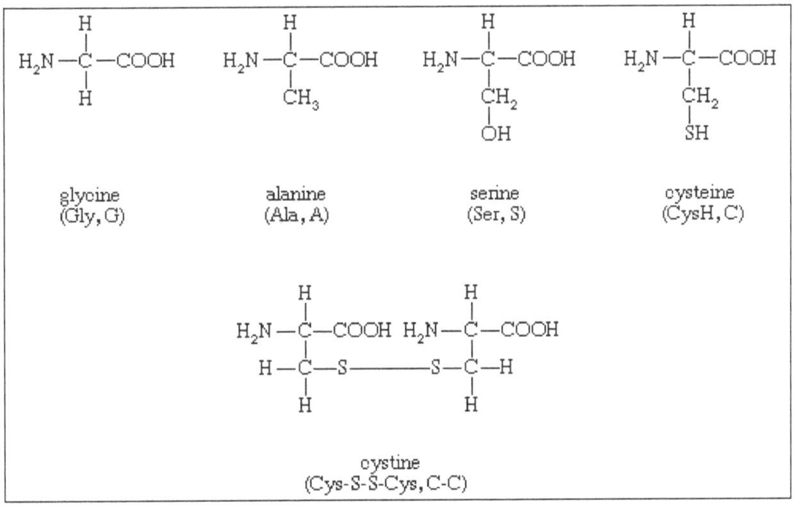

Four amino acids, each consisting of four carbon atoms, occur in proteins; they are aspartic acid, asparagine, threonine, and methionine. Aspartic acid and asparagine, which occur in large amounts, can be synthesized by animals. Threonine and methionine cannot be synthesized and thus are essential amino acids; i.e., they must be supplied in the diet. Most proteins contain only small amounts of methionine.

Proteins also contain an amino acid with five carbon atoms (glutamic acid) and a secondary amine (in proline), which is a structure with the amino group ($-NH_2$) bonded to the alkyl side chain, forming a ring. Glutamic acid and aspartic acid are dicarboxylic acids; that is, they have two carboxyl groups ($-COOH$).

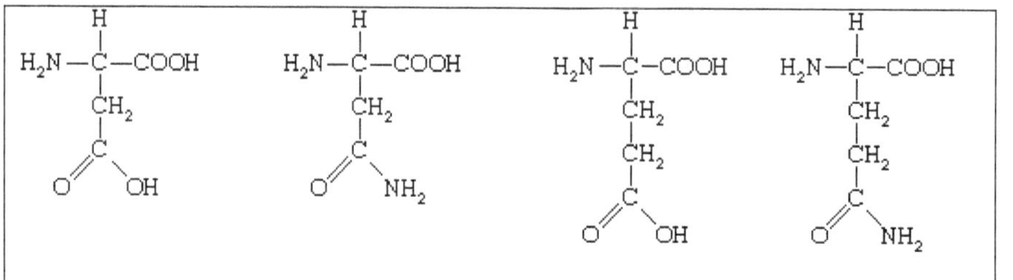

Glutamine is similar to asparagine in that both are the amides of their corresponding dicarboxylic acid forms; i.e., they have an amide group ($-CONH_2$) in place of the carboxyl ($-COOH$) of the side chain. Glutamic acid and glutamine are abundant in most proteins; e.g., in plant proteins they sometimes comprise more than one-third of the amino acids present. Both glutamic acid and glutamine can be synthesized by animals.

Table: Amino acid content of some proteins.

	Proteins					
amino acid	alpha-casein	gliadin	edestin	collagen (ox hide)	keratin (wool)	myosin
lysine	60.9	4.45	19.9	27.4	6.2	85
histidine	18.7	11.7	18.6	4.5	19.7	15
arginine	24.7	15.7	99.2	47.1	56.9	41
aspartic acid**	63.1	10.1	99.4	51.9	51.5	85
threonine	41.2	17.6	31.2	19.3	55.9	41
serine	63.1	46.7	55.7	41.0	79.5	41
glutamic acid**	153.1	311.0	144.9	76.2	99.0	155
proline	71.3	117.8	32.9	125.2	58.3	22
glycine	37.3	—	68.0	354.6	78.0	39
alanine	41.5	23.9	57.7	115.7	43.8	78
half-cystine	3.6	21.3	10.9	0.0	105.0	86
valine	53.8	22.7	54.6	21.4	46.6	42
methionine	16.8	11.3	16.4	6.5	4.0	22
isoleucine	48.8	90.8***	41.9	14.5	29.0	42
leucine	60.3		60.0	28.2	59.9	79
tyrosine	44.7	17.7	26.9	5.5	28.7	18
phenylalanine	27.9	39.0	38.4	13.9	22.4	27
tryptophan	7.8	3.2	6.6	0.0	9.6	—
hydroxyproline	0.0	0.0	0.0	97.5	12.2	—
hydroxylysine	—	—	—	8.0	1.2	—
total	839	765	883	1,058	863	832
average residual weight	119	131	113	95	117	120

*Number of gram molecules of amino acid per 100,000 grams of protein.
**The values for aspartic acid and glutamic acid include asparagine and glutamine, respectively.
***Isoleucine plus leucine.

The amino acids proline and hydroxyproline occur in large amounts in collagen, the protein of the connective tissue of animals. Proline and hydroxyproline lack free amino ($-NH_2$) groups because the amino group is enclosed in a ring structure with the side chain; they thus cannot exist in a zwitterion form. Although the nitrogen-containing group (>NH) of these amino acids can form a peptide bond with the carboxyl group of another amino acid, the bond so formed gives rise to a kink in the peptide chain; i.e., the ring structure alters the regular bond angle of normal peptide bonds.

Proteins usually are almost neutral molecules; that is, they have neither acidic nor basic properties. This means that the acidic carboxyl ($-COO)^-$ groups of aspartic and glutamic acid are about equal in number to the amino acids with basic side chains. Three such basic amino acids, each containing six carbon atoms, occur in proteins. The one with the simplest structure, lysine, is synthesized by plants but not by animals. Even some plants have a low lysine content. Arginine is found in all proteins; it occurs in particularly high amounts in the strongly basic protamines (simple proteins composed of relatively few amino acids) of fish sperm. The third basic amino acid is histidine. Both arginine and histidine can be synthesized by animals. Histidine is a weaker base than either lysine or arginine. The imidazole ring, a five-membered ring structure containing two nitrogen atoms in the side chain of histidine, acts as a buffer (i.e., a stabilizer of hydrogen ion concentration) by binding hydrogen ions (H^+) to the nitrogen atoms of the imidazole ring.

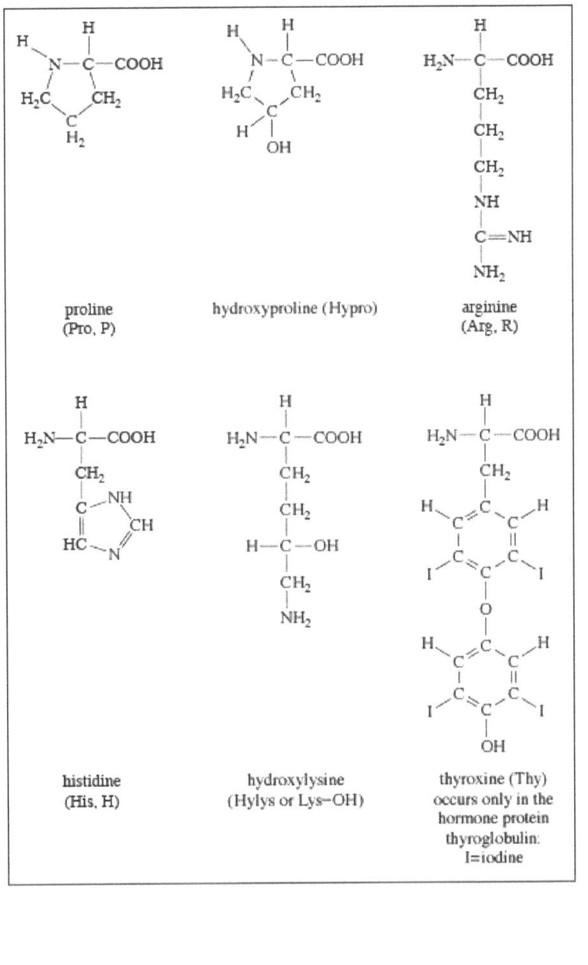

proline (Pro, P) — hydroxyproline (Hypro) — arginine (Arg, R)

histidine (His, H) — hydroxylysine (Hylys or Lys–OH) — thyroxine (Thy) occurs only in the hormone protein thyroglobulin: I=iodine

The remaining amino acids—phenylalanine, tyrosine, and tryptophan—have in common an aromatic structure; i.e., a benzene ring is present. These three amino acids are essential, and, while animals cannot synthesize the benzene ring itself, they can convert phenylalanine to tyrosine.

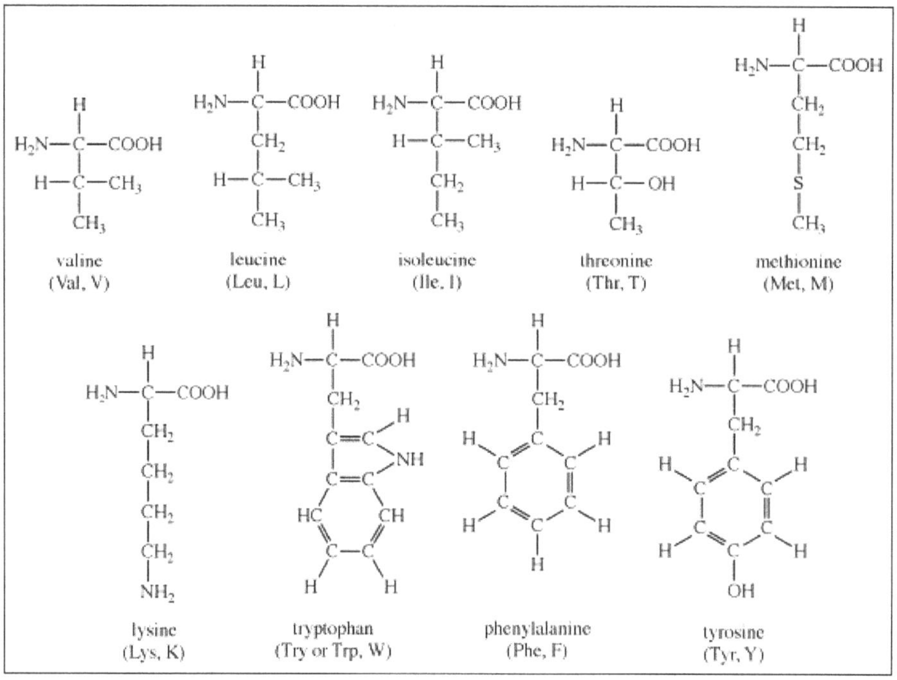

Because these amino acids contain benzene rings, they can absorb ultraviolet light at wavelengths between 270 and 290 nanometres (nm; 1 nanometre = 10−9 metre = 10 angstrom units). Phenylalanine absorbs very little ultraviolet light; tyrosine and tryptophan, however, absorb it strongly and are responsible for the absorption band most proteins exhibit at 280–290 nanometres. This absorption is often used to determine the quantity of protein present in protein samples.

However, other amino acids occur in proteins in small amounts. For example, the collagen found in connective tissue contains, in addition to hydroxyproline, small amounts of hydroxylysine. Other proteins contain some monomethyl-, dimethyl-, or trimethyllysine—i.e., lysine derivatives containing one, two, or three methyl groups ($-CH_3$). The amount of these unusual amino acids in proteins, however, rarely exceeds 1 or 2 percent of the total amino acids.

Physicochemical Properties of the Amino Acids

The physicochemical properties of a protein are determined by the analogous properties of the amino acids in it. The α-carbon atom of all amino acids, with the exception of glycine, is asymmetric; this means that four different chemical entities (atoms or groups of atoms) are attached to it. As a result, each of the amino acids, except glycine, can exist in two different spatial, or geometric, arrangements (i.e., isomers), which are mirror images akin to right and left hands.

These isomers exhibit the property of optical rotation. Optical rotation is the rotation of the plane of polarized light, which is composed of light waves that vibrate in one plane, or direction, only.

Solutions of substances that rotate the plane of polarization are said to be optically active, and the degree of rotation is called the optical rotation of the solution. The direction in which the light is rotated is generally designed as plus, or d, for dextrorotatory (to the right), or as minus, or l, for levorotatory (to the left). Some amino acids are dextrorotatory, others are levorotatory. With the exception of a few small proteins (peptides) that occur in bacteria, the amino acids that occur in proteins are L-amino acids.

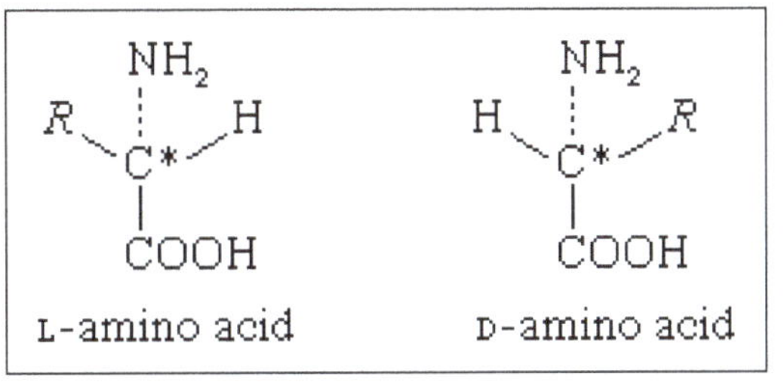

In bacteria, D-alanine and some other D-amino acids have been found as components of gramicidin and bacitracin. These peptides are toxic to other bacteria and are used in medicine as antibiotics. The D-alanine has also been found in some peptides of bacterial membranes.

In contrast to most organic acids and amines, the amino acids are insoluble in organic solvents. In aqueous solutions they are dipolar ions (zwitterions, or hybrid ions) that react with strong acids or bases in a way that leads to the neutralization of the negatively or positively charged ends, respectively. Because of their reactions with strong acids and strong bases, the amino acids act as buffers—stabilizers of hydrogen ion (H^+) or hydroxide ion (OH^-) concentrations. In fact, glycine is frequently used as a buffer in the pH range from 1 to 3 (acid solutions) and from 9 to 12 (basic solutions). In acid solutions, glycine has a positive charge and therefore migrates to the cathode (negative electrode of a direct-current electrical circuit with terminals in the solution). Its charge, however, is negative in alkaline solutions, in which it migrates to the anode (positive electrode). At pH 6.1 glycine does not migrate, because each molecule has one positive and one negative charge. The pH at which an amino acid does not migrate in an electrical field is called the isoelectric point. Most of the monoamino acids (i.e., those with only one amino group) have isoelectric points similar to that of glycine. The isoelectric points of aspartic and glutamic acids, however, are close to pH_3, and those of histidine, lysine, and arginine are at pH 7.6, 9.7, and 10.8, respectively.

Amino Acid Sequence in Protein Molecules

Since each protein molecule consists of a long chain of amino acid residues, linked to each other by peptide bonds, the hydrolytic cleavage of all peptide bonds is a prerequisite for the quantitative determination of the amino acid residues. Hydrolysis is most frequently accomplished by boiling the protein with concentrated hydrochloric acid. The quantitative determination of the amino acids is based on the discovery that amino acids can be separated from each other by chromatography on filter paper and made visible by spraying the paper with ninhydrin. The amino acids of the protein hydrolysate are separated from each other by passing the hydrolysate through a column of adsorbents, which adsorb the amino acids with different affinities and, on washing the column with buffer solutions, release them in a definite order. The amount of each of the amino acids can be determined by the intensity of the colour reaction with ninhydrin.

To obtain information about the sequence of the amino acid residues in the protein, the protein is degraded stepwise, one amino acid being split off in each step. This is accomplished by coupling the free α-amino group ($-NH_2$) of the N-terminal amino acid with phenyl isothiocyanate; subsequent mild hydrolysis does not affect the peptide bonds. The procedure, called the Edman degradation, can be applied repeatedly; it thus reveals the sequence of the amino acids in the peptide chain.

Unavoidable small losses that occur during each step make it impossible to determine the sequence of more than about 30 to 50 amino acids by this procedure. For this reason the protein is usually first hydrolyzed by exposure to the enzyme trypsin, which cleaves only peptide bonds formed by the carboxyl groups of lysine and arginine. The Edman degradation is then applied to each of the few resulting peptides produced by the action of trypsin. Further information can be gained by hydrolyzing another portion of the protein with another enzyme, for instance with chymotrypsin, which splits predominantly peptide bonds formed by the amino acids tyrosine, phenylalanine, and tryptophan. The combination of results obtained with two or more different proteolytic (protein degrading) enzymes was first applied by English biochemist Frederick Sanger, and it enabled him to elucidate the amino acid sequence of insulin. The amino acid sequences of many other proteins subsequently were determined in the same manner.

Protein Structure

Protein structure depends on its amino acid sequence and local, low-energy chemical bonds between atoms in both the polypeptide backbone and in amino acid side chains. Protein structure plays a key role in its function; if a protein loses its shape at any structural level, it may no longer be functional. The term, structure, when used in relation to proteins, takes on a much more complex meaning than it does for small molecules. Proteins are macromolecules and have four different levels of structure – primary, secondary, tertiary and quaternary. Primary structure is the amino acid sequence. Secondary structure is local interactions between stretches of a polypeptide chain and includes α-helix and β-pleated sheet structures. Tertiary structure is the overall the three-dimension folding driven largely by interactions between R groups. Quarternary structures are the orientation and arrangement of subunits in a multi-subunit protein.

The building blocks of proteins are amino acids, which are small organic molecules that consist of an alpha (central) carbon atom linked to an amino group, a carboxyl group, a hydrogen atom, and a variable component called a side chain. Within a protein, multiple amino acids are linked together by peptide bonds, thereby forming a long chain. Peptide bonds are formed by a biochemical reaction that extracts a water molecule as it joins the amino group of one amino acid to the carboxyl group of a neighboring amino acid. The linear sequence of amino acids within a protein is considered the primary structure of the protein.

Proteins are built from a set of only twenty amino acids, each of which has a unique side chain. The side chains of amino acids have different chemistries. The largest group of amino acids has nonpolar side chains. Several other amino acids have side chains with positive or negative charges, while others have polar but uncharged side chains. The chemistry of amino acid side chains is critical to protein structure because these side chains can bond with one another to hold a length of protein in a certain shape or conformation. Charged amino acid side chains can form ionic bonds, and polar amino acids are capable of forming hydrogen bonds. Hydrophobic side chains interact with each other via weak van der Waals interactions. The vast majority of bonds formed by these side chains are noncovalent. In fact, cysteines are the only amino acids capable of forming covalent bonds, which they do with their particular side chains. Because of side chain interactions, the sequence and location of amino acids in a particular protein guides where the bends and folds occur in that protein.

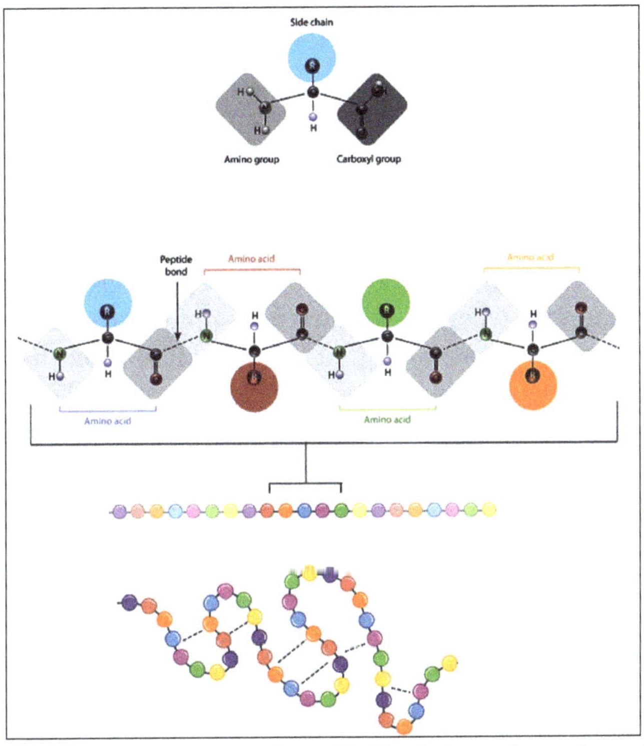

Figure: The relationship between amino acid side chains and protein conformation.

The defining feature of an amino acid is its side chain (at top, blue circle; below, all colored circles). When connected together by a series of peptide bonds, amino acids form a polypeptide, another word for protein. The polypeptide will then fold into a specific conformation depending on the interactions (dashed lines) between its amino acid side chains.

Protein Primary Structure

The primary structure of a protein is defined as the sequence of amino acids of which it is composed. This sequence ultimately determines the shape that the protein adopts, according to the spatial limitations on the arrangement of the atoms in the protein, the chemical properties of the component amino acid residues, and the protein's environment.

The peptide bonds that link amino acid residues in a polypeptide are formed in a condensation reaction between the acidic carboxyl group of one amino acid and the basic amino group of another amino acid. In the context of a peptide, the amide group (CO–NH) is referred to as the peptide group.

Crucial to an understanding of protein structure is knowledge of the structure of the peptide bond. Linus Pauling, in the 1930s, used X-ray diffraction to examine the nature of the peptide bond formed between two amino acids. He reported that the peptide group (CO–NH) has a rigid planar structure.

This structure is due to interactions between electrons of the double bond of the carbonyl group and those of the C–N bond such that the latter acquires partial (about 40%) double-bond properties.

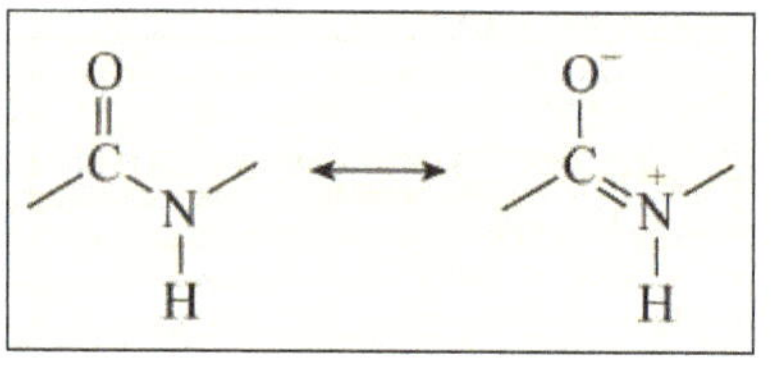

Figure: Resonance interactions between electrons in the C=O bond and the C–N bond of the peptide group mean that there is 'sharing' of electrons between these bonds. Note the charges on the N and O atoms.

This effect is an example of resonance which can be thought of as a sharing of electrons between bonds. Since single bonds between two atoms are longer than double bonds between the same two atoms, the lengths of the C–N and C=O bonds in the peptide group differ from those observed for these bonds in other contexts where resonance does not occur. Thus the partial double bond character of C–N in the peptide group means that this bond is shorter than would be predicted for a C–N single bond, whilst the C=O bond, having a partial single bond character due to resonance, is longer than would be predicted for a C=O double bond. The bond lengths in the peptide group are indicated in Figure. Compare the C–N bond of the peptide group with that between N and Cα (the C atom to which the amino group and carboxyl group are attached).

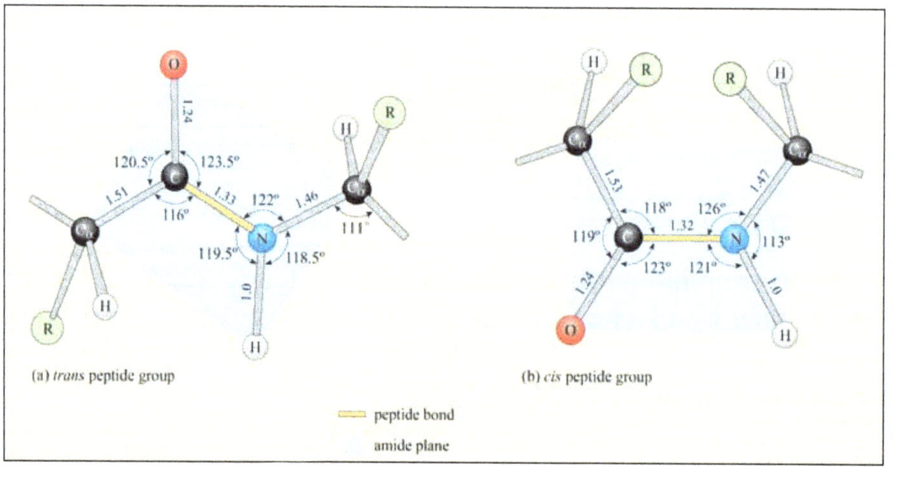

Figure shows the average dimensions in angstroms, Å (10 Å = 1 nm = 1000 pm), and degrees, of the planar peptide group in (a) the trans conformation and (b) the cis conformation. Note: in this and other representations of the configuration of polypeptides, there is no indication of bond order (i.e. single or double bonds).

There are two possible conformations of the planar peptide bond: in the trans peptide group, the Cα atoms are on opposite sides of the peptide bond and in the cis peptide group, the Cα atoms are on the same side of the peptide bond.

Generally speaking, peptide bonds are in the transconformation. However, cis forms can occur in peptide bonds that precede a proline residue. In such cases, the cis form is more stable than usual since the proline side-chain offers less of a hindrance. Nonetheless, cis peptide bonds occur only in approximately 10% of instances of peptide bonds preceding proline residues.

Bearing in mind the planar nature of the peptide group, a polypeptide chain can be seen to have a backbone that consists of a series of rigid planar peptide groups linked by the Cα atoms. Figure shows part of a polypeptide with two planar peptide groups in the transconformation. Note that though rotation is not permitted about the peptide bonds, there is potential for rotation around the Cα–N and Cα–C bonds. The angles of rotation, termed torsion angles, about these bonds specify the conformation of a polypeptide backbone. The torsion angles about the Cα–N and Cα–C bonds are referred to as ϕ (phi) and ψ. (psi), respectively and they are defined as 180° when the polypeptide is in the extended planar conformation, as illustrated in Figure below.

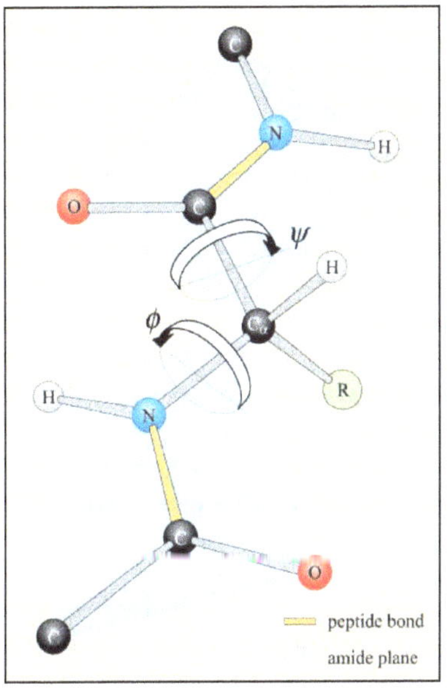

Figure shows a portion of a polypeptide showing two planar peptide groups (shaded areas) joined at the Cα atom. There is little freedom of movement within the peptide group, but rotation of the peptide groups about the Cα–N and Cα–C bonds can occur and is defined by the torsion angles ϕ and ψ respectively. In the extended conformation shown here, both these angles are defined as 180°. By convention, ϕ and ϕ increase with clockwise rotation of the peptide group when viewed from Cα, as indicated by the arrows.

As a result of these steric constraints, only certain values of ϕ and ψ, and hence conformations of the peptide, are permitted whilst others are not. It is possible to calculate these permitted values for a given residue in the context of a polypeptide. This calculation is performed by first determining the distances between all the non-bonding atoms in two neighbouring peptide groups (such as that in figure) at all the possible values of ϕ and ψ. It is most readily done for a polypeptide containing just one kind of amino acid. A conformational plot of ϕ against ψ for a particular residue is known as a Ramachandran plot (after its inventor, G. N. Ramachandran). Such a plot allows us to identify those conformations (i.e. for a particular value of ϕ and ψ) that are sterically favourable or unfavourable, according to the following criteria:

- Where there is no conflict between the van der Waals radii of non-bonding atoms, a conformation is 'allowed'. These conformations lie in the blue areas in Figure below.
- Conformations requiring interatomic distances at the limit of that which is permissible are defined as 'outer limit' conformations. They lie in the green areas in Figure below.

- Theoretical conformations that require any two non-bonding atoms to be closer to each other than their van der Waals radii allow are sterically 'forbidden'. These lie in the white areas in Figure below.

Notice that the values of ϕ and ψ in Figure range from −180° to +180°. Turning the peptide group through 360° will of course bring it back to its starting position, and −180° and +180° correspond to the same position. Thus the green strip at the bottom left corner of the plot in Figure is contiguous with the field at the top left corner.

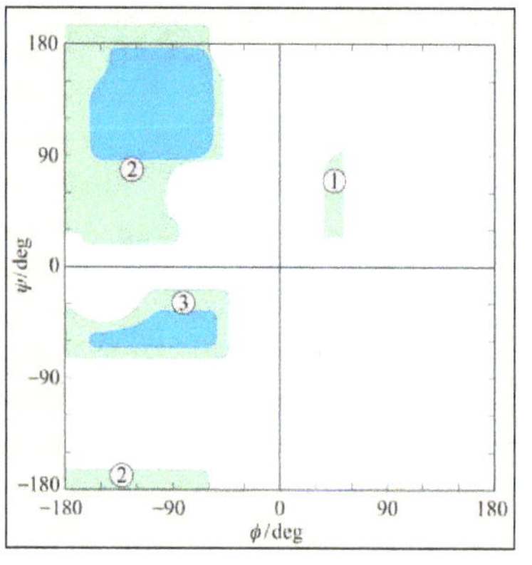

Figure shows a Ramachandran plot showing the sterically allowed ϕ and ψ angles for a peptide chain containing only l-alanine residues (poly-l-alanine). 'Allowed' conformations are within the blue or green areas, with those that are 'outer limit' conformations in the latter. The remaining conformations, in the white area, are 'forbidden'. Note that there are only three discrete regions (numbered 1–3) corresponding to allowed conformations.

Ramachandran plots can be constructed for polymers of each of the 20 amino acids. It is significant to note that the Ramachandran plots for many amino acid residues are generally very similar, having only three regions with favourable or tolerated conformations (labelled 1–3 in the plot for poly-l-alanine in Figure above). Differences do occur, however. For instance, where the side-chain (R in Figure) is branched near Cα, as in the case of threonine, it occupies more space close to the peptide backbone and restricts the approach of atoms in the neighbouring peptide groups. As a result, allowed conformations (ϕ and ψ angles) are more restricted for polypeptides of branched amino acids.

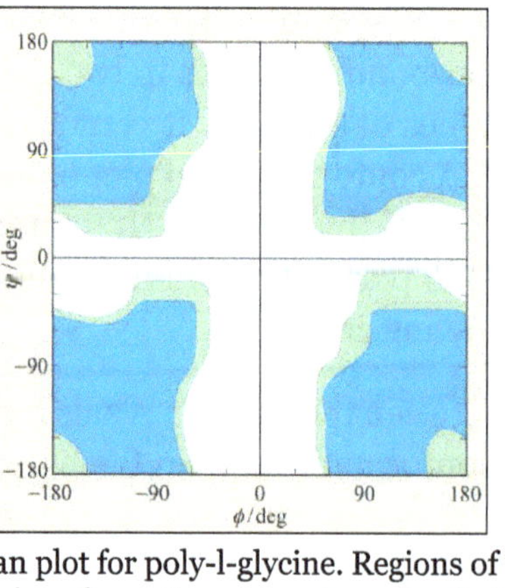

Figure: A Ramachandran plot for poly-l-glycine. Regions of 'allowed', 'outer limit' and 'forbidden' conformations are coloured as in Figure above.

The Ramachandran plots in Figures have been generated for, respectively, l-alanine and l-glycine on the basis of allowed and outer limit distances for interatomic contacts, determined from known values for van der Waals radii of the atoms.

Table: Van der Waals distances for interatomic contacts.

Contact type	Normally allowed / Å	Outer limit / Å
H···H	2.0	1.9
H···O	2.4	2.2
H···N	2.4	2.2
H···C	2.4	2.2
O···O	2.7	2.6
O···N	2.7	2.6
O···C	2.8	2.7
N···N	2.7	2.6
N···C	2.9	2.8
C···C	3.0	2.9
C···CH_2	3.2	3.0
CH_2···CH_2	3.2	3.0

They are therefore predictive rather than actual conformational plots. We can, of course, use X-ray diffraction to determine experimentally the 'real' values of ϕ and ψ for residues in a polypeptide. In Figure below, the ϕ and ψ values for all the residues (with the exception of glycine and proline) in a number of different structures have been determined by high-resolution X-ray diffraction and plotted on a Ramachandran plot. We can see that there is a striking correspondence between predicted and actual conformations. Notice, however, that there are some residues whose conformations map to the 'forbidden' areas. Most of these residues map in the region between 'allowed' regions 2 and 3, around $\psi = 0$.

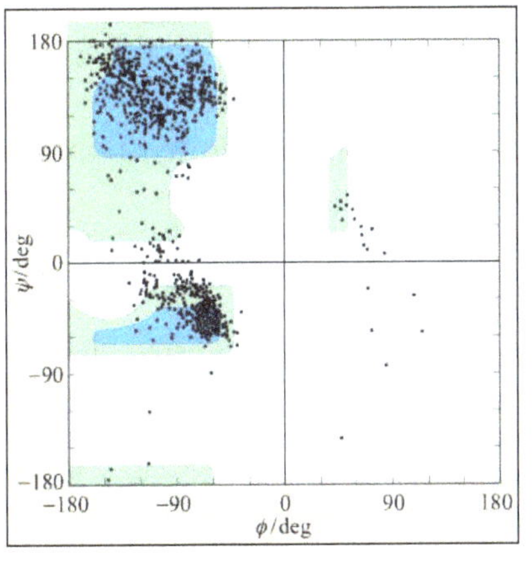

Figure shows based on X-ray diffraction data for a number of polypeptide structures, the ϕ and ψ values for all amino acid residues (with the exception of glycine and proline) have been superimposed on a predicted Ramachandran plot of 'allowed' and 'outer limit' conformations. Predictions were based on van der Waals distances for interatomic contacts as described in the text. Notice that the majority of actual values correspond to predicted permissible conformations.

The conflict associated with these conformations can be accommodated by a small degree of twisting of the peptide bond. Thus, in such conformations the peptide group is twisted out of its usual planar conformation.

A limited number of 'forbidden' conformations of particular residues can be tolerated in a polypeptide if the adopted conformation, as a whole, is energetically favourable. A polypeptide will tend to fold such that it adopts the most stable conformation. In this conformation, the polypeptide minimises its free energy.

Protein Secondary Structure

Secondary structures are those repetitive structures involving H bond between amide H and carbonyl O in- the main chain. These include alpha helices, beta strands (sheets) and reverse turns.

Alpha Structure

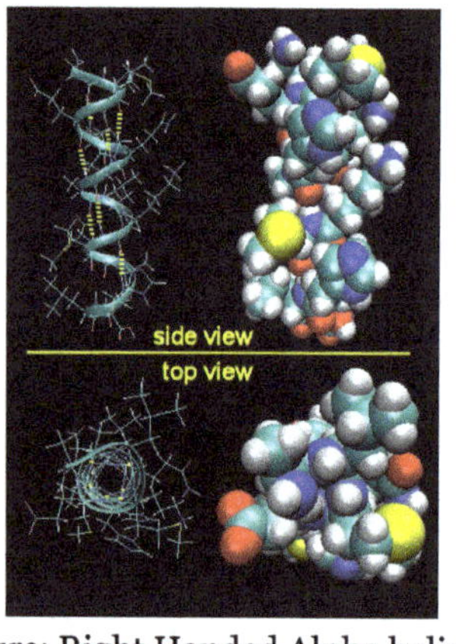

Figure: Right Handed Alpha helices.

These helices are formed when the carbonyl O of the i th amino acid H bonds to the amide H of the i th +4 aa (4 amino acids away). The phi/psi angles for those amino acids in the alpha helix are -57,-47, which emphasizes the regular repeating nature of the structure. It can also be characterized by n (the number of amino acid units/turn = 3.6) and pitch (the helix rise/turn = 5.4 angstroms). The alpha helix is more compact than the fully extended polypeptide chain with phi/psi angles of 180°.

- In proteins, the average number of amino acids in a helix is 11, which gives 3 turns.
- The left-handed alpha helix, although allowed from inspections of a Ramachandran plot, is never observed, since the side chains are too close to the backbone.
- The core of the helix is packed tightly. There are not holes or pores in the helix.
- All the R-groups extend backward and away from the helix axis.
- Some amino acids are more commonly found in alpha helices than other. Amino acids can be divided into two kinds, those with branches at the beta C and those with none. Consider first those that aren't branched. Gly is too conformationally flexible to be found with high frequency in alpha helices, while Pro is too rigid. The amino acids with side chains that can

H-bond (Ser, Asp, and Asn) and aren't too long appear to act as competitors of main chain H bond donor and acceptors, and destabilize alpha helices. The rest with no branches at the beta C can form helices. Those with branches at the beta carbon (Val, Ile) destabilize the alpha helix due to steric interactions of the bulky side chains with the helix backbone. Remember left-handed alpha helices are not found in nature for similar reasons.

- Alpha keratins, the major component of hair, skin, fur, beaks, and fingernails, are almost all alpha helix.

Helices

There are other kinds of helices that can occur. These include a 310 helix and a \(\pi\) helix, which are stabilized by H-bonds between the amide NH and carbonyl O of residues (i, i+3) and (i, i+5), respectively. Likewise, they have 3 and 4.3 residues/turn, respectively, and a rise per residue of 6 and 4.7 angstrom, respectively. These structures are much rarer than right handed alpha helices.

Helix Type	H bond btw ith and ith+X AA, where X =	Residue/turn	Rise (Angstrom)/ turn
310	3	3	6
α	4	3.6	5.4
π	5	4.3	4.7

Beta Structure

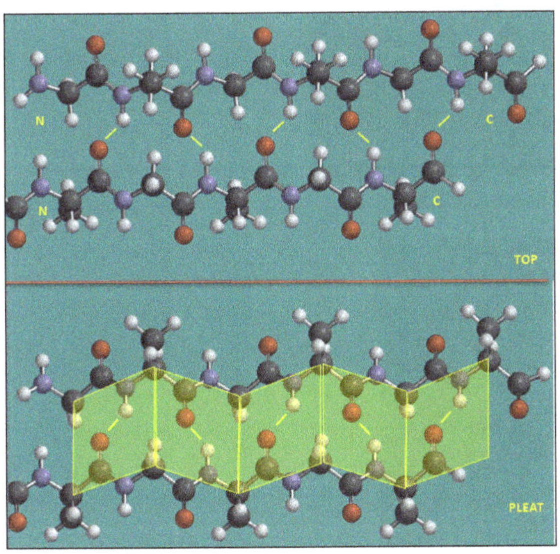

Figure: Parallel beta strands.

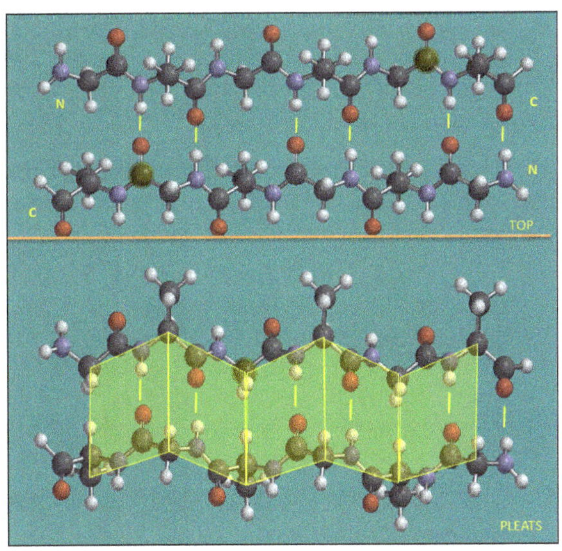

Figure: Antiparallel beta strands.

Beta Structure: Parallel and antiparallel beta strands are much more extended than alpha helices (phi/psi of -57,-47) but not as extended as a fully extended polypeptide chain (with phi/psi angles of +/- 180). The beta sheets are not quit so extended (parallel -119, +113 ; antiparallel, -139, +135), and can be envisioned as rippled sheets. They can be visualized by laying thin, pleated strips of paper side by side to make a "pleated sheet" of paper. Each strip of paper can be pictured as a single peptide strand in which the peptide backbone makes a zig-zag along the strip, with the alpha carbons lying at the folds of the pleats. Each single strand of the beta-sheet can be pictured as a twofold helix, i.e. a helix with 2 residues/turn. The arrangement of each successive peptide plane is pleated due to the tetrahedral nature of the alpha C. The H bonds are interstrand, not intrastrand as in the alpha helix.

In a parallel beta sheet structure, the optimal H bond pattern leads to a less extended structure (phi/psi of -119, +113) than the optimal arrangement of the H bonds in the antiparallel structure (phi/psi of -139, +135). Also the H bonds in the parallel sheet are bent significantly. (i.e. the carbonyl O on one strand is not exactly opposite the amide H on the adjacent strand, as it is in the antiparallel sheet.) Hence antiparallel beta strands are presumably more stable, even though both are abundantly found in nature. Short parallel beta sheets of 4 strands or less are not common, which might reflect their lower stability.

The side chains in the beta sheet are normal to the plane of the sheet, extending out from the plane on alternating sides. Parallel sheets characteristically distribute hydrophobic side chains on both side of the sheet, while antiparallel sheets are usually arranged with all the hydrophobic residues on one side. This requires an alternation of hydrophilic and hydrophobic side chains in the primary sequence. Antiparallel sheets are found in silk with the sheets running parallel to the silk fibers. The following repeat is found in the primary sequence: (Ser-Gly-Ala-Gly)n), with Gly pointing out from one face, and Ser or Ala from the other. Beta strands have a tendency to twist in the right hand direction. This leads to important consequences in how the beta strands are connected.

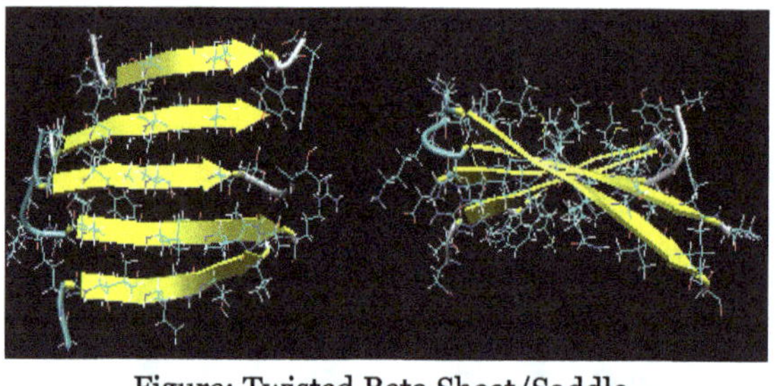

Figure: Twisted Beta Sheet/Saddle.

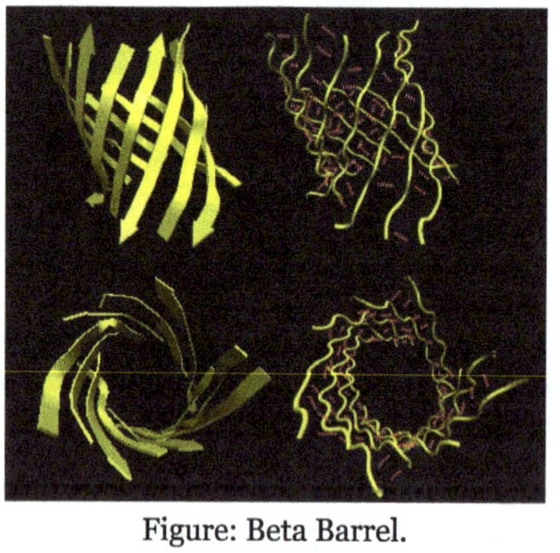

Figure: Beta Barrel.

- In parallel strands, right handed connectivity is common.
- In a protein with parallel strand in register, and given the inherent twist in the stands, the strands arrange in a way to have the H bonds stretched equally at the ends of the chains, giving rise to a twisted saddle shape.

- In a protein with parallel strand out of register, and given the inherent twist in the stands, the strands arrange in a way to have the H bonds stretched equally at the ends of the chains, giving rise to a beta barrel.

Reverse Turns

About 50% of the amino acids in a globular protein are in regular secondary structure (alpha or beta). The remaining amino acids are not less ordered, just less regular. An additional example of secondary structures is reverse turns (or beta-bends or beta turns). Reverse turns often connect successive antiparallel beta strands and are then called beta hairpins.

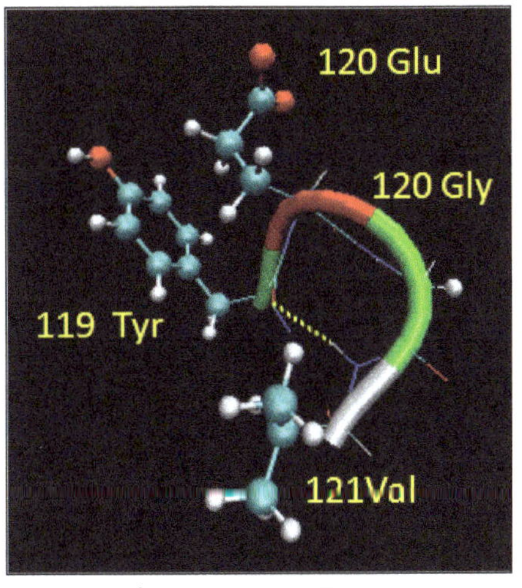

Figure: Reverse Turns.

They are almost always at the surface, and consist of 4 amino acids. There are two types. (I - f2 = -60, y2=-30; f3 = -90, y3 = 0; II - f2 = -60, y2=120; f3 = 90, y3 = 0) Residue 2 of both is often Pro. (Why?) Both have an H bond between the carbonyl O of the i th a.a and the amide H of the i th+3 aa (three amino acids away). In the type 2, the O of residue 2 crowds the beta C of residue 3, so aa2 is usually Gly. Why? Those amino acids which destabilize alpha helices are often found in beta sheets, since the side chains project out of the plan which holds the main chain.

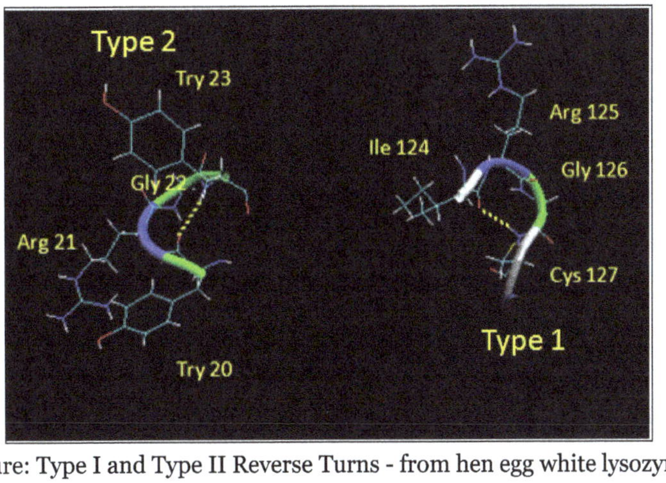

Figure: Type I and Type II Reverse Turns - from hen egg white lysozyme.

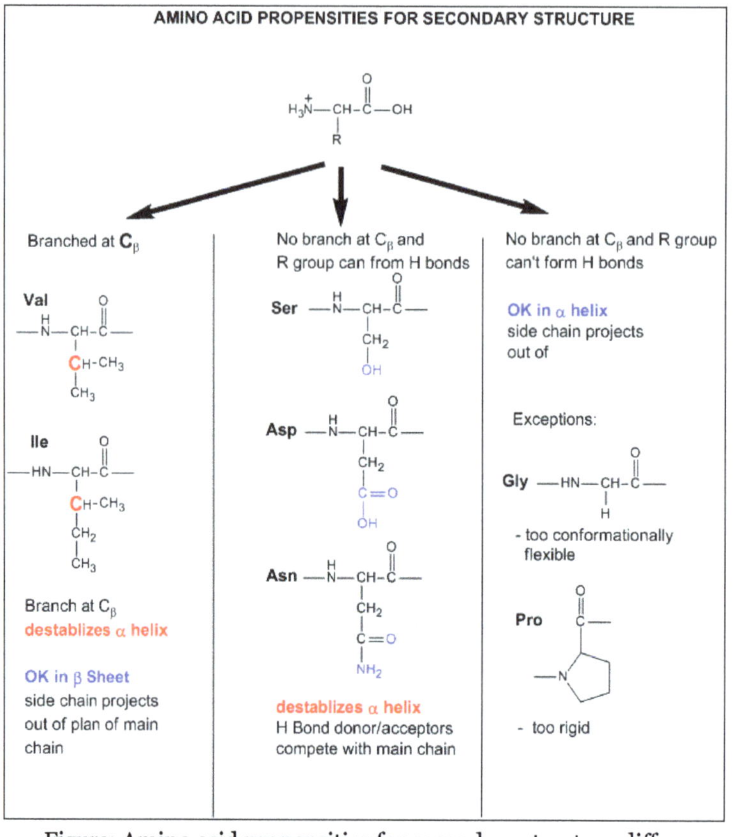

Figure: Amino acid propensities for secondary structure differ.

Protein Tertiary Structure

The tertiary structure is the structure at which polypeptide chains become functional. At this level, every protein has a specific three-dimensional shape and presents functional groups on its outer surface, allowing it to interact with other molecules, and giving it its unique function. The arrangement is made with the help of chaperones, which move the protein chain around, bringing different groups on the chain closer together in order to help them form bonds. These amino acids interacting are usually far away from each other on the chain.

The primary structure of a protein, which is the simple chain of amino acids held together by peptide bonds, is what determines the higher-order or secondary and tertiary, structures by dictating the folding of the chain. Every amino acid has a unique side chain, or R-group, which is what, gives amino acids their distinct properties.

When a protein, such as an enzyme, loses its tertiary structure, it can no longer do its job because it has become denatured and has lost its biological function. This usually happens at temperatures that are too high for the protein molecule. However, once temperatures are returned to normal, the tertiary structure can be achieved again. This suggests that it is the primary structure that is the most important for determining the more complex folding.

Tertiary Structure Interactions

The following are the main interactions that make up the tertiary structures of proteins.

They guide the bending and twisting that help the protein molecule achieve a stable state. We can observe interactions that are covalent, where pairs of electrons are shared between atoms, or non-covalent, where pairs of electrons are not shared between atoms. Recall that the breaking down of these bonds can lead to the denaturation of the protein.

Hydrophobic Interactions

These non-covalent bonds are the most important factor and driving force in the formation of the tertiary structure. If we place hydrophobic (water-hating) molecules in water, these molecules will aggregate together and form large chunks of hydrophobic molecules. Since some R-groups are hydrophilic (water-loving) and others are hydrophobic, all the amino acids containing the hydrophilic side chains, such as isoleucine, will be found on the surface of the protein, while the amino acids that have hydrophobic side chains, like alanine, will aggregate together at the center of the protein. Therefore, a protein that forms in water, as most of them do, will have a hydrophobic core and a hydrophilic surface. This is crucial in determining what the tertiary structure will look like.

Disulfide Bridges

These are very strong covalent bonds found between cysteine residues that are in close proximity in space. The bonds form between the sulfur groups on the different cysteine residues, as shown below.

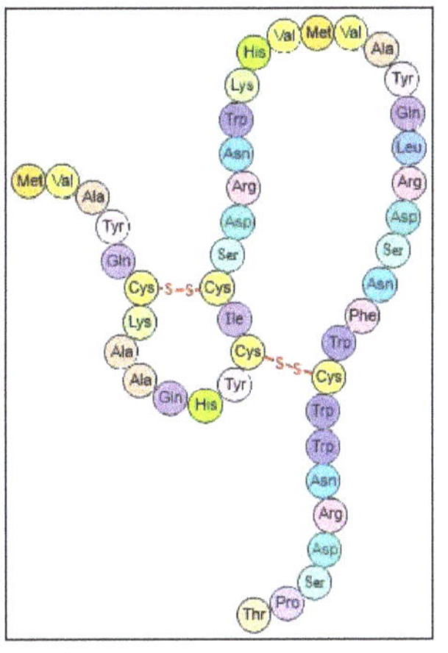

Ionic Bonds

Some amino acids contain side chains that carry positive or negative charges. If an amino acid with a positive charge comes close enough to an amino acid that carries a negative charge, they can from a bond that helps to stabilize the protein molecule.

Hydrogen Bonds

We can observe these bonds between water molecules in the solution and the hydrophilic amino acid side chains on the surface of the molecule. Hydrogen bonds also occur between polar side chains and help in stabilizing the tertiary structure.

Types of Tertiary Structures

Globular Proteins

Most proteins fall into this category. Globular proteins form a compact ball shape, where hydrophobic amino acids are found in the center of the structure and hydrophilic amino acids are found on the surface, forming a molecule that is soluble in water. Many globular proteins have domains, which are locally folded parts of the tertiary structure, ranging from 50 amino acids to 350 amino acids. One domain can be found in more than one protein if the proteins have similar functions, and a protein with multiple functions can have more than one domain, each playing a specific role. An example of globular proteins is the enzymes found within our cells.

Fibrous Proteins

Fibrous proteins are made of fibers often consisting of repeated sequences of amino acids, resulting in a highly ordered, elongated molecule. They include cartilage, which provides structural support, and are insoluble in water.

Protein Quaternary Structure

Quaternary structure refers to the spatial arrangement of subunits and the nature of their interactions. The simplest sort of quaternary structure is a dimer, consisting of two identical subunits. This organization is present in the DNA-binding protein Cro found in a bacterial virus called λ. More complicated quaternary structures also are common. More than one type of subunit can be present, often in variable numbers. For example, human hemoglobin, the oxygen-carrying protein in blood, consists of two subunits of one type (designated α) and two subunits of another type (designated β), as illustrated in Figure. Thus, the hemoglobin molecule exists as a α2β2 tetramer. Subtle changes in the arrangement of subunits within the hemoglobin molecule allow it to carry oxygen from the lungs to tissues with great efficiency.

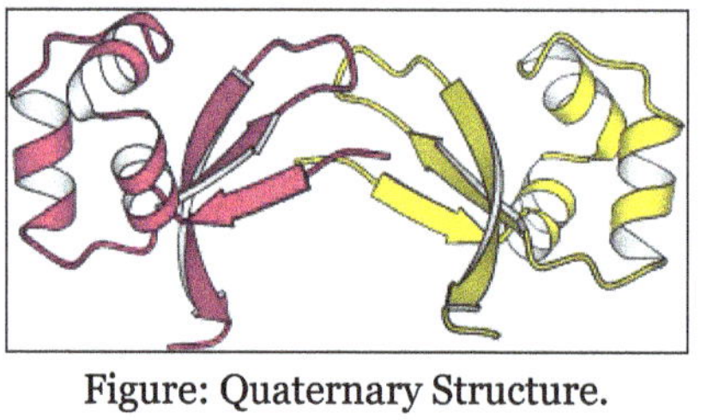

Figure: Quaternary Structure.

The Cro protein of bacteriophage λ is a dimer of identical subunits.

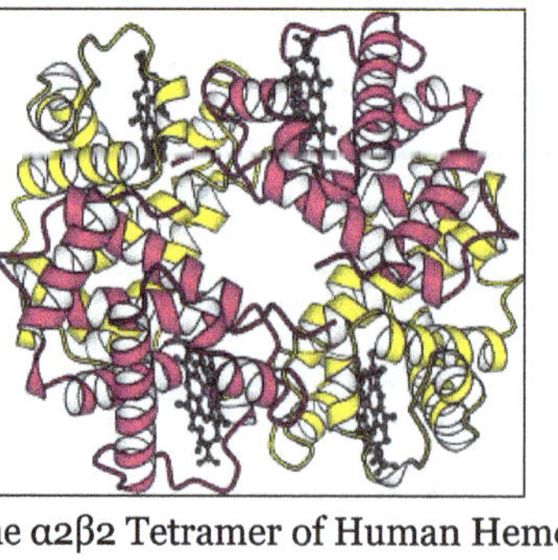

Figure: The α2β2 Tetramer of Human Hemoglobin.

The structure of the two identical α subunits (red) is similar to but not identical with that of the two identical β subunits (yellow). The molecule contains four heme groups.

Viruses make the most of a limited amount of genetic information by forming coats that use the same kind of subunit repetitively in a symmetric array. The coat of rhinovirus, the virus that causes the common cold, includes 60 copies each of four subunits. The subunits come together to form a nearly spherical shell that encloses the viral genome.

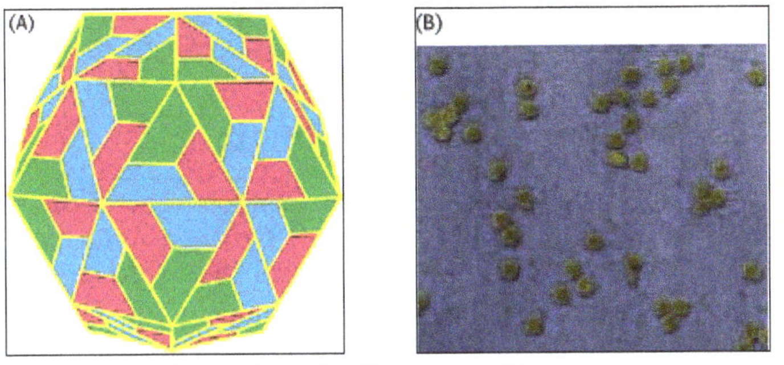

Figure: Complex Quaternary Structure.

Figure shows: The coat of rhinovirus comprises 60 copies of each of four subunits. (A) A schematic view depicting the three types of subunits (shown in red, blue, and green) visible from outside the virus. (B) An electron micrograph showing rhinovirus particles.

Protein Function

Each type of protein consists of a precise sequence of amino acids that allows it to fold up into a particular three-dimensional shape, or conformation. But proteins are not rigid lumps of material. They can have precisely engineered moving parts whose mechanical actions are coupled to chemical events. It is this coupling of chemistry and movement that gives proteins the extraordinary capabilities that underlie the dynamic processes in living cells.

All Proteins Bind to Other Molecules

The biological properties of a protein molecule depend on its physical interaction with other molecules. Thus, antibodies attach to viruses or bacteria to mark them for destruction, the enzyme hexokinase binds glucose and ATP so as to catalyze a reaction between them, actin molecules bind to each other to assemble into actin filaments, and so on. Indeed, all proteins stick, or bind, to other molecules. In some cases, this binding is very tight; in others, it is weak and short-lived. But the binding always shows great specificity, in the sense that each protein molecule can usually bind just one or a few molecules out of the many thousands of different types it encounters. The substance that is bound by the protein—no matter whether it is an ion, a small molecule, or a macromolecule— is referred to as a ligand for that protein.

The ability of a protein to bind selectively and with high affinity to a ligand depends on the formation of a set of weak, noncovalent bonds—hydrogen bonds, ionic bonds, and van der Waals attractions—plus favorable hydrophobic interactions. Because each individual bond is weak, an effective binding interaction requires that many weak bonds be formed simultaneously. This is possible only if the surface contours of the ligand molecule fit very closely to the protein, matching it like a hand in a glove.

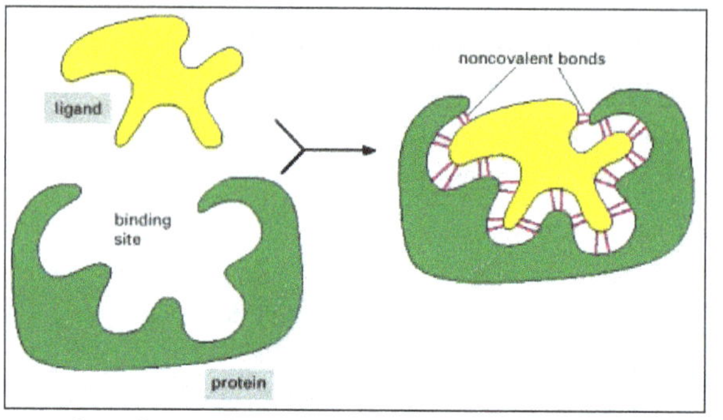

Figure: The selective binding of a protein to another molecule.

Many weak bonds are needed to enable a protein to bind tightly to a second molecule, which is called a ligand for the protein. A ligand must therefore fit precisely into a protein's binding site, like a hand into a glove, so that a large number of noncovalent bonds can be formed between the protein and the ligand.

The region of a protein that associates with a ligand, known as the ligand's binding site, usually consists of a cavity in the protein surface formed by a particular arrangement of amino acids. These amino acids can belong to different portions of the polypeptide chain that are brought together when the protein folds. Separate regions of the protein surface generally provide binding sites for different ligands, allowing the protein's activity to be regulated. And other parts of the protein can serve as a handle to place the protein in a particular location in the cell—an example is the SH2 domain, which is often used to move a protein containing it to sites in the plasma membrane in response to particular signals.

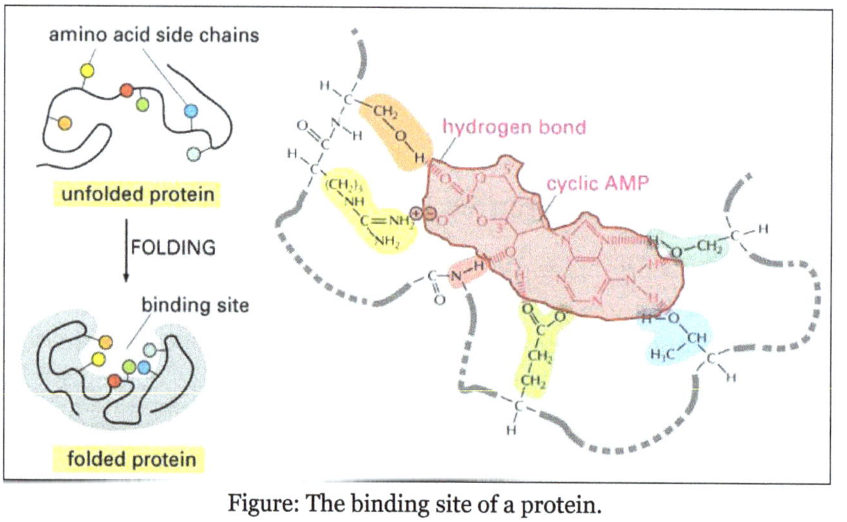

Figure: The binding site of a protein.

Figure shows (A) The folding of the polypeptide chain typically creates a crevice or cavity on the protein surface. This crevice contains a set of amino acid side chains disposed in such a way that they can make noncovalent bonds only with certain ligands. (B) A close-up of an actual binding site showing the hydrogen bonds and ionic interactions formed between a protein and its ligand (in this example, cyclic AMP is the bound ligand).

Although the atoms buried in the interior of the protein have no direct contact with the ligand, they provide an essential scaffold that gives the surface its contours and chemical properties. Even small changes to the amino acids in the interior of a protein molecule can change its three-dimensional shape enough to destroy a binding site on the surface.

The details of a protein's conformation determine its chemistry Proteins have impressive chemical capabilities because the neighboring chemical groups on their surface often interact in ways that enhance the chemical reactivity of amino acid side chains. These interactions fall into two main categories.

First, neighboring parts of the polypeptide chain may interact in a way that restricts the access of water molecules to a ligand binding site. Because water molecules tend to form hydrogen bonds, they can compete with ligands for sites on the protein surface. The tightness of hydrogen bonds (and ionic interactions) between proteins and their ligands is therefore greatly increased if water molecules are excluded. Initially, it is hard to imagine a mechanism that would exclude a molecule as small as water from a protein surface without affecting the access of the ligand itself. Because of the strong tendency of water molecules to form water–water hydrogen bonds, however, water molecules exist in a large hydrogen-bonded network. In effect, a ligand binding site can be kept dry because it is energetically unfavorable for individual water molecules to break away from this network, as they must do to reach into a crevice on a protein's surface.

Second, the clustering of neighboring polar amino acid side chains can alter their reactivity. If a number of negatively charged side chains are forced together against their mutual repulsion by the way the protein folds, for example, the affinity of the site for a positively charged ion is greatly increased. In addition, when amino acid side chains interact with one another through hydrogen bonds, normally unreactive side groups (such as the –CH2OH on the serine) can become reactive, enabling them to enter into reactions that make or break selected covalent bonds.

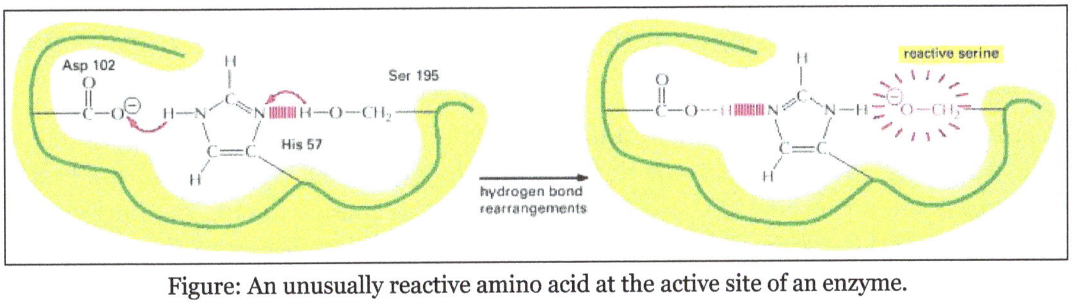

Figure: An unusually reactive amino acid at the active site of an enzyme.

This example is the "catalytic triad" found in chymotrypsin, elastase, and other serine proteases. The aspartic acid side chain (Asp 102) induces the histidine (His 57) to remove the proton from serine 195. This activates the serine to form a covalent bond with the enzyme substrate, hydrolyzing a peptide bond.

The surface of each protein molecule therefore has a unique chemical reactivity that depends not only on which amino acid side chains are exposed, but also on their exact orientation relative to one another. For this reason, even two slightly different conformations of the same protein molecule may differ greatly in their chemistry. Sequence Comparisons Between Protein Family Members Highlight Crucial Ligand Binding Sites.

Many of the domains in proteins can be grouped into families that show clear evidence of their evolution from a common ancestor, and genome sequences reveal large numbers of proteins that contain one or more common domains. The three-dimensional structures of the members of the same domain family are remarkably similar. For example, even when the amino acid sequence identity falls to 25%, the backbone atoms in a domain have been found to follow a common protein fold within 0.2 nanometers (2 Å).

These facts allow a method called "evolutionary tracing" to be used to identify those sites in a protein domain that are the most crucial to the domain's function. For this purpose, those amino acids that are unchanged, or nearly unchanged, in all of the known protein family members are mapped onto a structural model of the three-dimensional structure of one family member. When this is done, the most invariant positions often form one or more clusters on the protein surface, as illustrated in Figure for the SH2 domain. These clusters generally correspond to ligand binding sites.

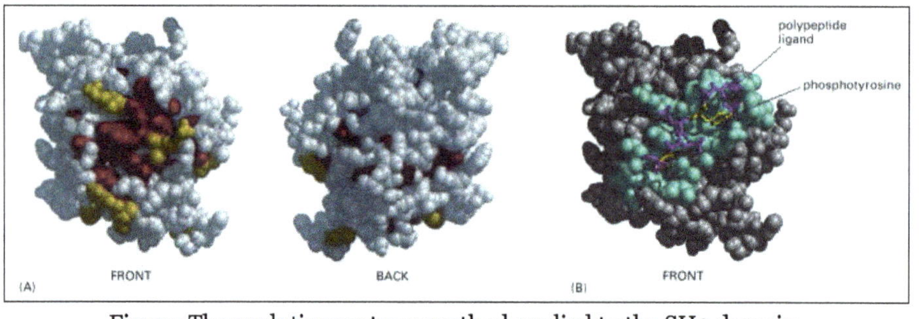

Figure: The evolutionary trace method applied to the SH2 domain.

Figure shows (A) Front and back views of a space-filling model of the SH2 domain, with evolutionarily conserved amino acids on the protein surface colored yellow, and those more toward the protein interior colored red. (B) The structure of the SH2 domain with its bound polypeptide. Here, those amino acids located within 0.4 nm of the bound ligand are colored blue. The two key amino acids of the ligand are yellow, and the others are purple.

The SH2 domain is a module that functions in protein–protein interactions. It binds the protein containing it to a second protein that contains a phosphorylated tyrosine side chain in a specific amino acid sequence context, as shown in figure B. The amino acids located at the binding site for the phosphorylated polypeptide have been the slowest to change during the long evolutionary process that produced the large SH2 family of peptide recognition domains. Because mutation is a random process, this result is attributed to the preferential elimination during evolution of all organisms whose SH2 domains became altered in a way that inactivated the SH2-binding site, thereby destroying the function of the SH2 domain.

In this era of extensive genome sequencing, many new protein families have been discovered whose functions are unknown. By identifying the critical binding sites on a three-dimensional structure determined for one family member, the above method of evolutionary tracing is being used to help determine the functions of such proteins.

Proteins Bind to Other Proteins through Several Types of Interfaces

Proteins can bind to other proteins in at least three ways. In many cases, a portion of the surface of one protein contacts an extended loop of polypeptide chain (a "string") on a second protein. Such a surface–string interaction, for example, allows the SH2 domain to recognize a phosphorylated polypeptide as a loop on a second protein, as just described, and it also enables a protein kinase to recognize the proteins that it will phosphorylate.

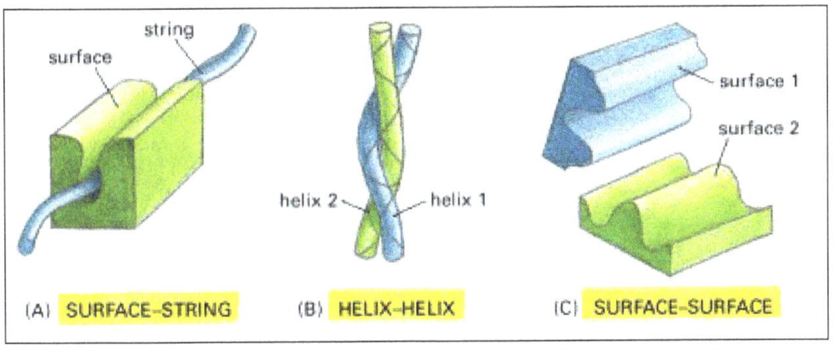

Figure: Three ways in which two proteins can bind to each other.

Figure shows Only the interacting parts of the two proteins are shown. (A) A rigid surface on one protein can bind to an extended loop of polypeptide chain (a "string") on a second protein. (B) Two α helices can bind together to form a coiled-coil. (C) Two complementary rigid surfaces often link two proteins together.

A second type of protein–protein interface is formed when two α helices, one from each protein, pair together to form a coiled-coil. This type of protein interface is found in several families of gene regulatory proteins.

The most common way for proteins to interact, however, is by the precise matching of one rigid surface with that of another. Such interactions can be very tight, since a large number of weak bonds can form between two surfaces that match well. For the same reason, such surface–surface interactions can be extremely specific, enabling a protein to select just one partner from the many thousands of different proteins found in a cell.

The Binding Sites of Antibodies are Especially Versatile

All proteins must bind to particular ligands to carry out their various functions. This capacity for tight selective binding is displayed to an extraordinary degree by the antibody family.

Antibodies, or immunoglobulins, are proteins produced by the immune system in response to foreign molecules, such as those on the surface of an invading microorganism. Each antibody binds to a particular target molecule extremely tightly, thereby either inactivating the target directly or marking it for destruction. An antibody recognizes its target (called an antigen) with remarkable specificity. Because there are potentially billions of different antigens we might encounter, we have to be able to produce billions of different antibodies.

Antibodies are Y-shaped molecules with two identical binding sites that are complementary to a small portion of the surface of the antigen molecule. A detailed examination of the antigen-binding sites of antibodies reveals that they are formed from several loops of polypeptide chain that protrude from the ends of a pair of closely juxtaposed protein domains. The enormous diversity of antigen-binding sites possessed by different antibodies is generated by changing only the length and amino acid sequence of these loops, without altering the basic protein structure.

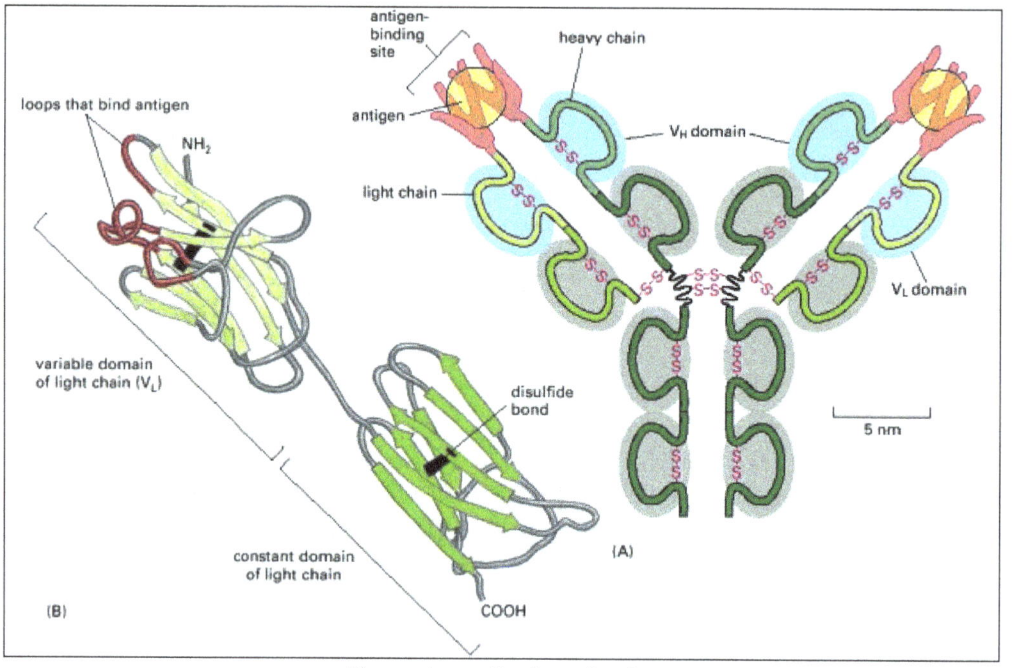

Figure: An antibody molecule.

Figure shows (A) A typical antibody molecule is Y-shaped and has two identical binding sites for its antigen, one on each arm of the Y. The protein is composed of four polypeptide chains (two identical heavy chains and two identical and smaller light chains) held together by disulfide bonds. Each chain is made up of several different immunoglobulin domains, here shaded either blue or gray. The antigen-binding site is formed where a heavy-chain variable domain (VH) and a light-chain variable domain (VL) come close together. These are the domains that differ most in their sequence and structure in different antibodies. (B) This ribbon model of a light chain shows the parts of the VL domain most closely involved in binding to the antigen in red. They contribute half of the fingerlike loops that fold around each of the antigen molecules in (A).

Loops of this kind are ideal for grasping other molecules. They allow a large number of chemical groups to surround a ligand so that the protein can link to it with many weak bonds. For this reason, loops are often used to form the ligand-binding sites in proteins.

Binding Strength is Measured by the Equilibrium Constant

Molecules in the cell encounter each other very frequently because of their continual random thermal movements. When colliding molecules have poorly matching surfaces, few noncovalent bonds form, and the two molecules dissociate as rapidly as they come together. At the other extreme, when many noncovalent bonds form, the association can persist for a very long time. Strong interactions occur in cells whenever a biological function requires that molecules remain associated for a long time—for example, when a group of RNA and protein molecules come together to make a subcellular structure such as a ribosome.

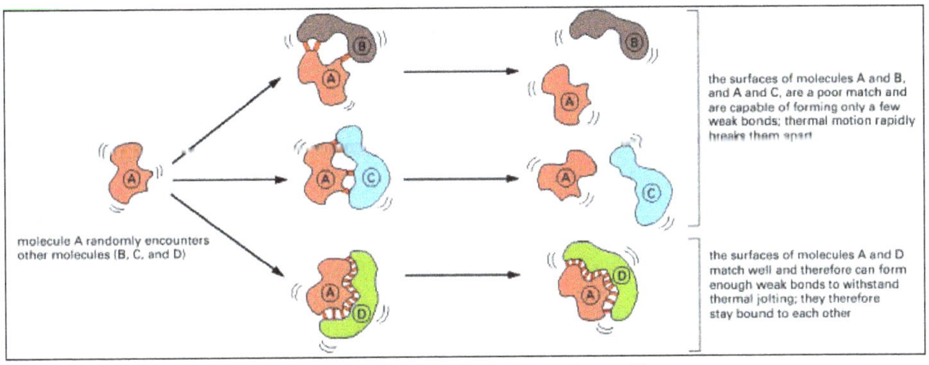

Figure: How noncovalent bonds mediate interactions between macromolecules?

The strength with which any two molecules bind to each other can be measured directly. As an example, imagine a situation in which a population of identical antibody molecules suddenly encounters a population of ligands diffusing in the fluid surrounding them. At frequent intervals, one of the ligand molecules will bump into the binding site of an antibody and form an antibody–ligand complex. The population of antibody–ligand complexes will therefore increase, but not without limit: over time, a second process, in which individual complexes break apart because of thermally induced motion, will become increasingly important. Eventually, any population of antibody molecules and ligands will reach a steady state, or equilibrium, in which the number of binding (association) events per second is precisely equal to the number of "unbinding" (dissociation) events.

From the concentrations of the ligand, antibody, and antibody–ligand complex at equilibrium, one can calculate a convenient measure—termed the equilibrium constant (K)—of the strength of binding. The equilibrium constant is greater the greater the binding strength and it is a direct measure of the free-energy difference between the bound and Free states. Even a change of a few noncovalent bonds can have a striking effect on a binding interaction, as shown by the example in Figure below.

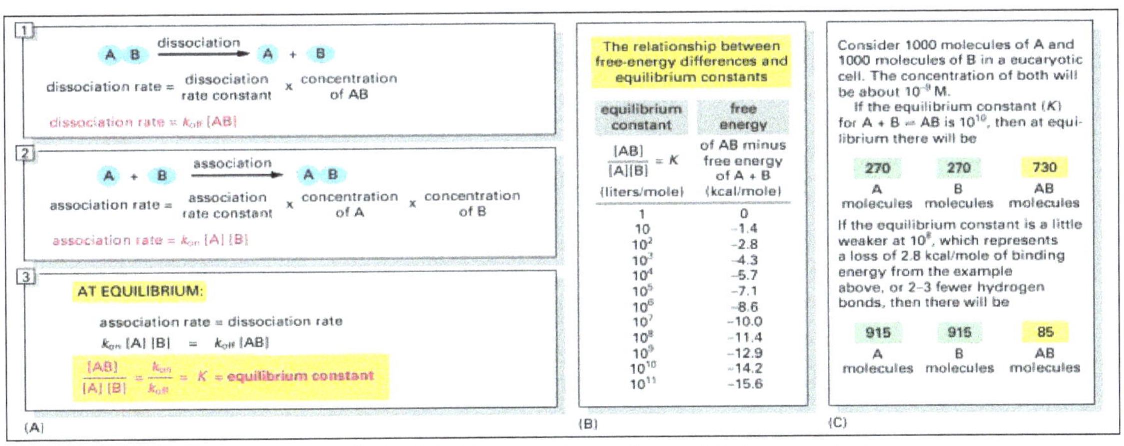

Figure: Relating binding energies to the equilibrium constant.

Figure shows (A) The equilibrium between molecules A and B and the complex AB is maintained by a balance between the two opposing reactions shown in panels 1 and 2. Molecules A and B must collide if they are to react, and the association rate is therefore proportional to the product of their individual concentrations [A] × [B]. (Square brackets indicate concentration.) As shown in panel

3, the ratio of the rate constants for the association and the dissociation reactions is equal to the equilibrium constant (K) for the reaction. (B) The equilibrium constant in panel 3 is that for the association reaction A + B ↔ AB, and the larger its value, the stronger the binding between A and B. Note that, for every 1.4 kcal/mole of free-energy drop, the equilibrium constant increases by a factor of 10. (C) An example of the dramatic effect that the presence or absence of a few weak bonds can have in a biological context.

The equilibrium constant here has units of liters/mole: for simple binding interactions it is also called the affinity constant or association constant, denoted Ka. The reciprocal of Ka is called the dissociation constant, Kd (in units of moles/liter).

Enzymes are Powerful and Highly Specific Catalysts

Many proteins can perform their function simply by binding to another molecule. An actin molecule, for example, need only associate with other actin molecules to form a filament. There are other proteins, however, for which ligand binding is only a necessary first step in their function. This is the case for the large and very important class of proteins called enzymes. Enzymes are remarkable molecules that determine all the chemical transformations that make and break covalent bonds in cells. They bind to one or more ligands, called substrates, and convert them into one or more chemically modified products, doing this over and over again with amazing rapidity. Enzymes speed up reactions, often by a factor of a million or more, without themselves being changed—that is, they act as catalysts that permit cells to make or break covalent bonds in a controlled way. It is the catalysis of organized sets of chemical reactions by enzymes that creates and maintains the cell, making life possible.

Enzymes can be grouped into functional classes that perform similar chemical reactions. Each type of enzyme within such a class is highly specific, catalyzing only a single type of reaction. Thus, hexokinase adds a phosphate group to d-glucose but ignores its optical isomer l-glucose; the blood-clotting enzyme thrombin cuts one type of blood protein between a particular arginine and its adjacent glycine and nowhere else, and so on. Enzymes work in teams, with the product of one enzyme becoming the substrate for the next. The result is an elaborate network of metabolic pathways that provides the cell with energy and generates the many large and small molecules that the cell needs.

Table: Some Common Types of Enzymes.

ENZYME	REACTION CATALYZED
Hydrolases	General term for enzymes that catalyze a hydrolytic cleavage reaction.
Nucleases	Break down nucleic acids by hydrolyzing bonds between nucleotides.
Proteases	Break down proteins by hydrolyzing bonds between amino acids.
Synthases	General name used for enzymes that synthesize molecules in anabolic reactions by condensing two smaller molecules together.
Isomerases	Catalyze the rearrangement of bonds within a single molecule.
Polymerases	Catalyze polymerization reactions such as the synthesis of DNA and RNA.
Kinases	Catalyze the addition of phosphate groups to molecules. Protein kinases are an important group of kinases that attach phosphate groups to proteins.

Phosphatases	Catalyze the hydrolytic removal of a phosphate group from a molecule.
Oxido-Reductases	General name for enzymes that catalyze reactions in which one molecule is oxidized while the other is reduced. Enzymes of this type are often called oxidases, reductases, and dehydrogenases.
ATPases	Hydrolyze ATP: Many proteins with a wide range of roles have an energy-harnessing ATPase activity as part of their function, for example, motor proteins such as myosin and membrane transport proteins such as the sodium–potassium pump.

Enzyme names typically end in "-ase," with the exception of some enzymes, such as pepsin, trypsin, thrombin and lysozyme that were discovered and named before the convention became generally accepted at the end of the nineteenth century. The common name of an enzyme usually indicates the substrate and the nature of the reaction catalyzed. For example, citrate synthase catalyzes the synthesis of citrate by a reaction between acetyl CoA and oxaloacetate.

Substrate Binding is the First Step in Enzyme Catalysis

For a protein that catalyzes a chemical reaction (an enzyme), the binding of each substrate molecule to the protein is an essential prelude. In the simplest case, if we denote the enzyme by E, the substrate by S, and the product by P, the basic reaction path is E + S → ES → EP → E + P. From this reaction path, we see that there is a limit to the amount of substrate that a single enzyme molecule can process in a given time. If the concentration of substrate is increased, the rate at which product is formed also increases, up to a maximum value. At that point the enzyme molecule is saturated with substrate, and the rate of reaction (Vmax) depends only on how rapidly the enzyme can process the substrate molecule. This maximum rate divided by the enzyme concentration is called the turnover number. The turnover number is often about 1000 substrate molecules processed per second per enzyme molecule, although turnover numbers between 1 and 10,000 are known.

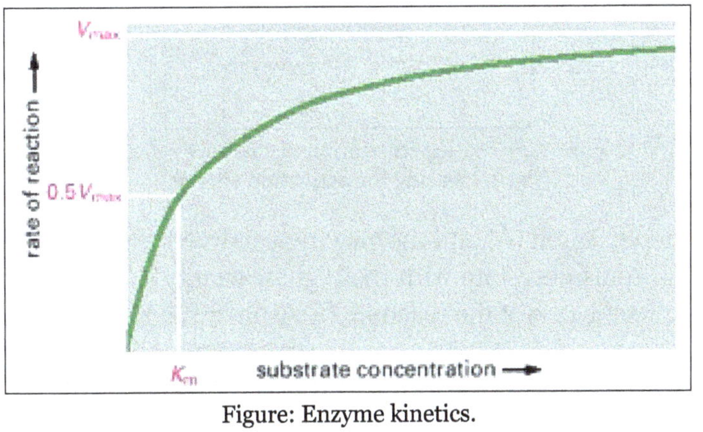

Figure: Enzyme kinetics.

The rate of an enzyme reaction (V) increases as the substrate concentration increases until a maximum value (V_{max}) is reached. At this point all substrate-binding sites on the enzyme molecules are fully occupied, and the rate of reaction is limited by the rate of the catalytic process on the enzyme surface. For most enzymes, the concentration of substrate at which the reaction rate is half-maximal (K_m) is a measure of how tightly the substrate is bound, with a large value of K_m corresponding to weak binding.

The other kinetic parameter frequently used to characterize an enzyme is its Km, the concentration of substrate that allows the reaction to proceed at one-half its maximum rate (0.5 Vmax). A low Km value means that the enzyme reaches its maximum catalytic rate at a low concentration of substrate and generally indicates that the enzyme binds to its substrate very tightly, whereas a high Km value corresponds to weak binding.

Enzymes Speed Reactions by Selectively Stabilizing Transition States

Extremely high rates of chemical reaction are achieved by enzymes—far higher than for any synthetic catalysts. This efficiency is attributable to several factors. The enzyme serves, first, to increase the local concentration of substrate molecules at the catalytic site and to hold all the appropriate atoms in the correct orientation for the reaction that is to follow. More importantly, however, some of the binding energy contributes directly to the catalysis. Substrate molecules must pass through a series of intermediate states of altered geometry and electron distribution before they form the ultimate products of the reaction. The free energy required to attain the most unstable transition state is called the activation energy for the reaction, and it is the major determinant of the reaction rate. Enzymes have a much higher affinity for the transition state of the substrate than they have for the stable form. Because this tight binding greatly lowers the energies of the transition state, the enzyme greatly accelerates a particular reaction by lowering the activation energy that is required.

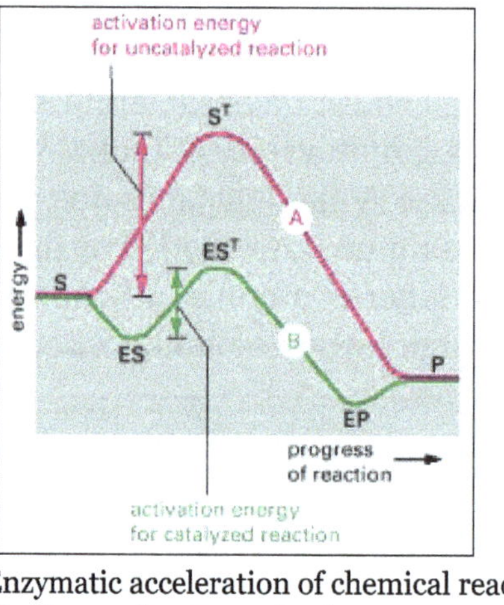

Figure: Enzymatic acceleration of chemical reactions|
by decreasing the activation energy.

Often both the uncatalyzed reaction (A) and the enzyme-catalyzed reaction (B) can go through several transition states. It is the transition state with the highest energy (ST and EST) that determines the activation energy and limits the rate of the reaction. (S = substrate; P = product of the reaction.)

A dramatic proof that stabilizing a transition state can greatly increase a reaction rate is provided by the intentional production of antibodies that act like enzymes. Consider, for example, the hydrolysis of an amide bond, which is similar to the peptide bond that joins two adjacent amino acids in a protein. In an aqueous solution, an amide bond hydrolyzes very slowly by the mechanism. In the central intermediate, or transition state, the carbonyl carbon is bonded to four atoms arranged at the corners of a tetrahedron. By generating monoclonal antibodies that bind tightly to a stable analog of this very unstable tetrahedral intermediate, an antibody that functions like an enzyme can be obtained. Because this catalytic antibody binds to and stabilizes the tetrahedral intermediate, it increases the spontaneous rate of amide-bond hydrolysis by more than 10,000-fold.

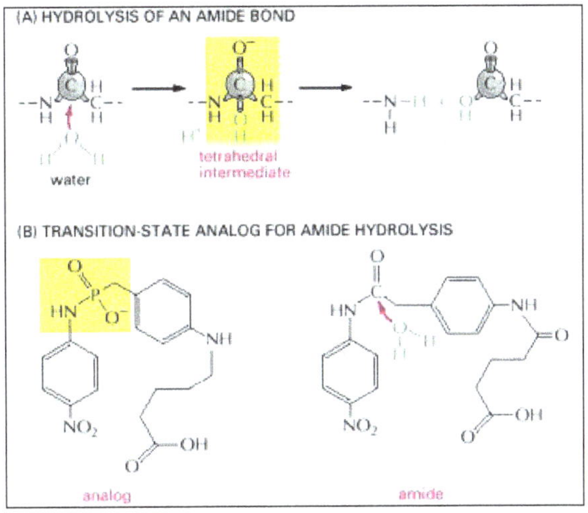

Figure: Catalytic antibodies.

The stabilization of a transition state by an antibody creates an enzyme. (A) The reaction path for the hydrolysis of an amide bond goes through a tetrahedral intermediate, the high-energy transition state for the reaction. (B) The molecule on the left was covalently linked to a protein and used as an antigen to generate an antibody that binds tightly to the region of the molecule shown in yellow. Because this antibody also bound tightly to the transition state in (A), it was found to function as an enzyme that efficiently catalyzed the hydrolysis of the amide bond in the molecule on the right.

Enzymes can use Simultaneous Acid and base Catalysis

Figures below compares the spontaneous reaction rates and the corresponding enzyme-catalyzed rates for five enzymes. Rate accelerations of 109–1023 are observed. Clearly, enzymes are much better catalysts than catalytic antibodies. Enzymes not only bind tightly to a transition state, they also contain precisely positioned atoms that alter the electron distributions in those atoms that participate directly in the making and breaking of covalent bonds. Peptide bonds, for example, can be hydrolyzed in the absence of an enzyme by exposing a polypeptide to either a strong acid or a strong base. Enzymes are unique, however, in being able to use acid and base catalysis simultaneously, since the acidic and basic residues required are prevented from combining with each other (as they would do in solution) by being tied to the rigid framework of the protein itself.

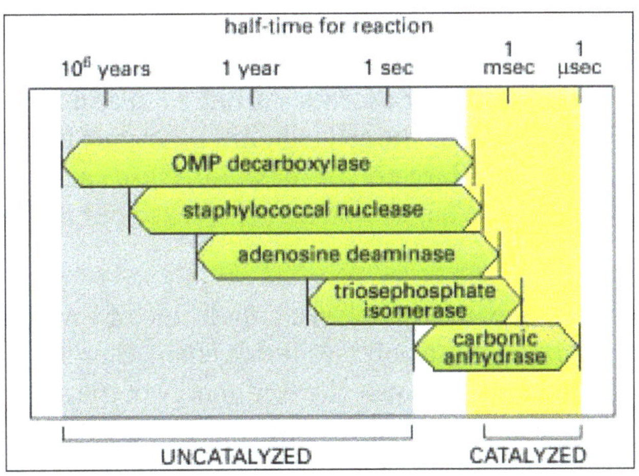

Figure: The rate accelerations caused by five different enzymes

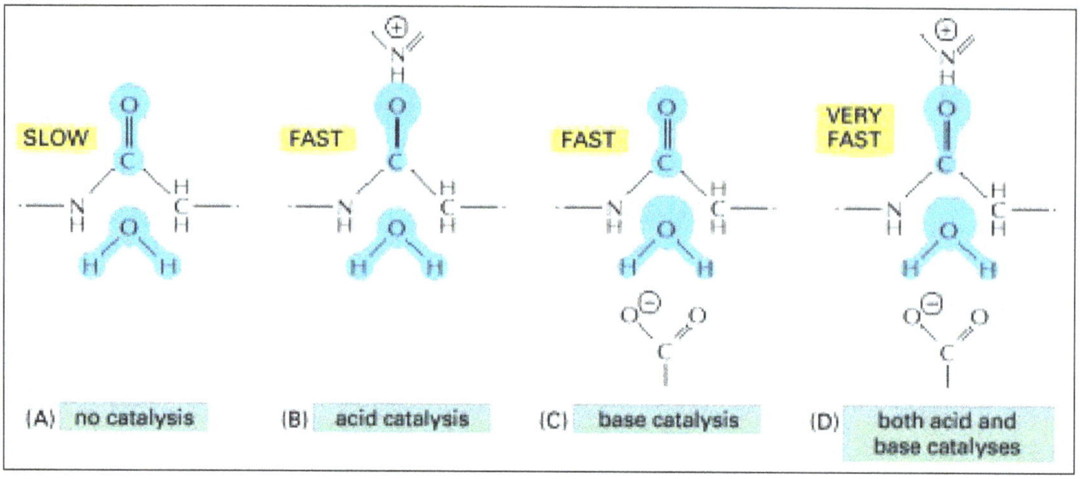

Figure: Acid catalysis and base catalysis.

Figure shows (A) The start of the uncatalyzed reaction shown in Figure is diagrammed, with blue indicating electron distribution in the water and carbonyl bonds. (B) An acid likes to donate a proton (H+) to other atoms. By pairing with the carbonyl oxygen, an acid causes electrons to move away from the carbonyl carbon, making this atom much more attractive to the electronegative oxygen of an attacking water molecule. (C) A base likes to take up H+. By pairing with a hydrogen of the attacking water molecule, a base causes electrons to move toward the water oxygen, making it a better attacking group for the carbonyl carbon. (D) By having appropriately positioned atoms on its surface, an enzyme can perform both acid catalysis and base catalysis at the same time.

The fit between an enzyme and its substrate needs to be precise. A small change introduced by genetic engineering in the active site of an enzyme can have a profound effect. Replacing a glutamic acid with an aspartic acid in one enzyme, for example, shifts the position of the catalytic carboxylate ion by only 1 Å (about the radius of a hydrogen atom); yet this is enough to decrease the activity of the enzyme a thousand fold.

Lysozyme Illustrate the Works of Enzyme

To demonstrate how enzymes catalyze chemical reactions, we shall use the example of an enzyme that acts as a natural antibiotic in egg white, saliva, tears, and other secretions. Lysozyme is an enzyme that catalyzes the cutting of polysaccharide chains in the cell walls of bacteria. Because the bacterial cell is under pressure from osmotic forces, cutting even a small number of polysaccharide chains causes the cell wall to rupture and the cell to burst. Lysozyme is a relatively small and stable protein that can be easily isolated in large quantities. For these reasons, it has been intensively studied, and it was the first enzyme to have its structure worked out in atomic detail by x-ray crystallography.

The reaction catalyzed by lysozyme is a hydrolysis: a molecule of water is added to a single bond between two adjacent sugar groups in the polysaccharide chain, thereby causing the bond to break. The reaction is energetically favorable because the free energy of the severed polysaccharide chain is lower than the free energy of the intact chain. However, the pure polysaccharide can sit for years in water without being hydrolyzed to any detectable degree. For a colliding water molecule to break a bond linking two sugars, the polysaccharide molecule has to be distorted into a particular shape—the transition state—in which the atoms around the bond have an altered geometry and electron distribution. Because of this distortion, large activation energy must be supplied through random collisions before the reaction can take place. In an aqueous solution at room temperature, the energy of collisions almost never exceeds the activation energy. Consequently, hydrolysis occurs extremely slowly, if at all.

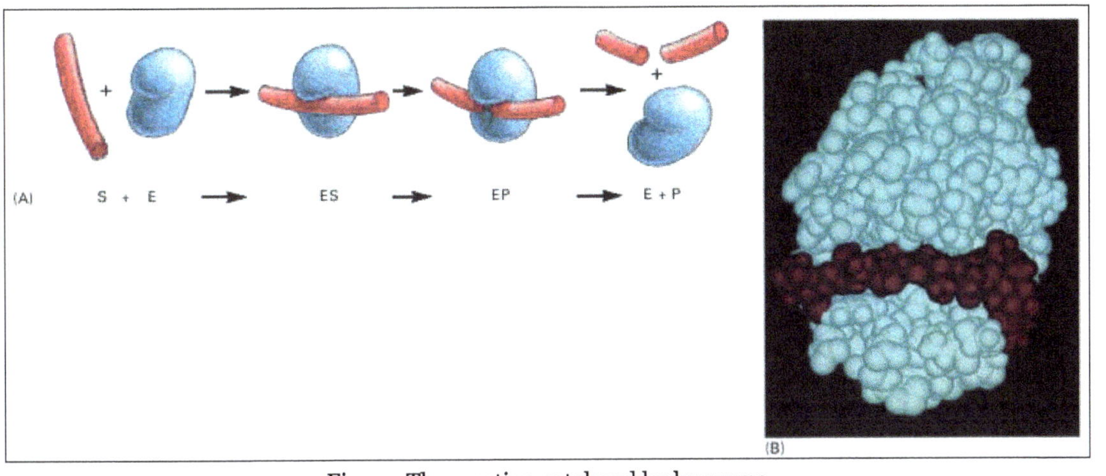

Figure: The reaction catalyzed by lysozyme.

Figure shows (A) The enzyme lysozyme (denoted as E) catalyzes the cutting of a polysaccharide chain, which is its substrate (S). The enzyme first binds to the chain to form an enzyme–substrate complex (ES) and then catalyzes the cleavage of a specific covalent bond in the backbone of the polysaccharide, forming an enzyme–product complex (EP) that rapidly dissociates. Release of the severed chain (the products P) leaves the enzyme free to act on another substrate molecule. (B) A space-filling model of the lysozyme molecule bound to a short length of polysaccharide chain before cleavage.

This situation is drastically changed when the polysaccharide binds to lysozyme. The active site of lysozyme, because its substrate is a polymer, is a long groove that holds six linked sugars at the same time. As soon as the polysaccharide binds to form an enzyme–substrate complex, the enzyme cuts the polysaccharide by adding a water molecule across one of its sugar–sugar bonds. The product chains are then quickly released, freeing the enzyme for further cycles of reaction.

The chemistry that underlies the binding of lysozyme to its substrate is the same as that for antibody binding to its antigen—the formation of multiple noncovalent bonds. However, lysozyme holds its polysaccharide substrate in a particular way, so that one of the two sugars involved in the bond to be broken is distorted from its normal, most stable conformation. The bond to be broken is also held close to two amino acids with acidic side chains (a glutamic acid and an aspartic acid) within the active site.

Conditions are thereby created in the microenvironment of the lysozyme active site that greatly reduce the activation energy necessary for the hydrolysis to take place. Figure below shows three central steps in this enzymatically catalyzed reaction.

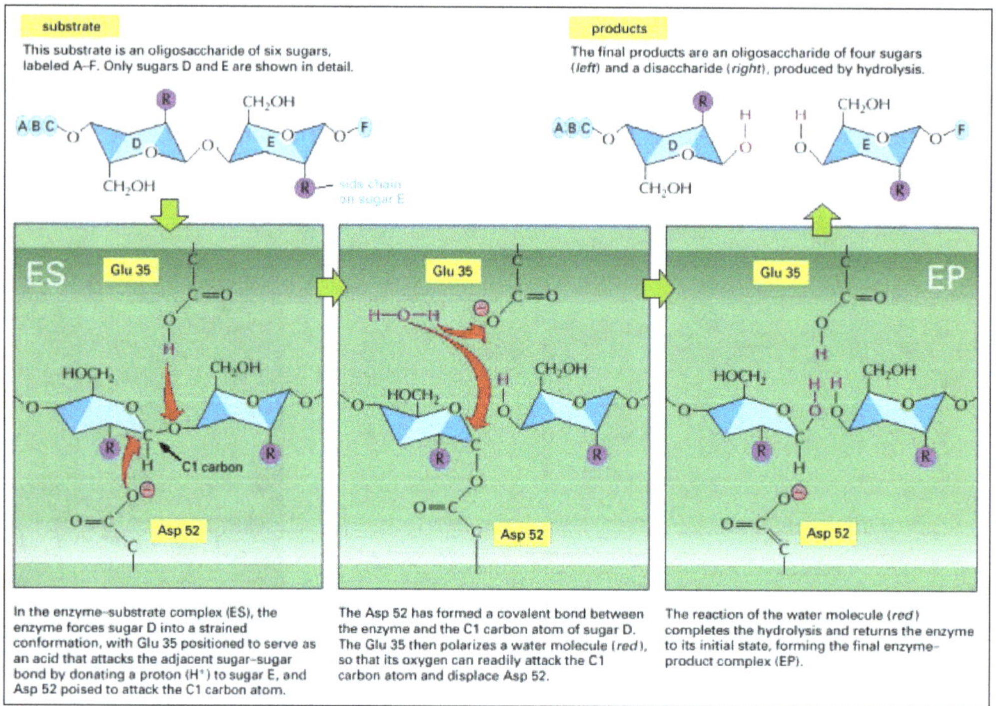

Figure: Events at the active site of lysozyme.

The top left and top right drawings depict the free substrate and the free products, respectively, whereas the other three drawings depict the sequential events at the enzyme active site. Note the change in the conformation of sugar D in the enzyme–substrate complex; this shape change stabilizes the oxocarbenium ion-like transition states required for formation and hydrolysis of the covalent intermediate shown in the middle panel.

- The enzyme stresses its bound substrate by bending some critical chemical bonds in one sugar, so that the shape of this sugar more closely resembles the shape of high-energy transition states formed during the reaction.
- The negatively charged aspartic acid reacts with the C1 carbon atom on the distorted sugar, breaking the sugar–sugar bond and leaving the aspartic acid side chain covalently linked to the site of bond cleavage.
- Aided by the negatively charged glutamic acid, a water molecule reacts with the C1 carbon atom, displacing the aspartic acid side chain and completing the process of hydrolysis.

The overall chemical reaction, from the initial binding of the polysaccharide on the surface of the enzyme through the final release of the severed chains, occurs many millions of times faster than it would in the absence of enzyme.

Similar mechanisms are used by other enzymes to lower activation energies and speed up the reactions they catalyze. In reactions involving two or more reactants, the active site also acts like a template, or mold, that brings the substrates together in the proper orientation for a reaction to occur between them. As we saw for lysozyme, the active site of an enzyme contains precisely positioned atoms that speed up a reaction by using charged groups to alter the distribution of electrons in the substrates. The binding to the enzyme also changes substrate shapes, bending bonds so as to drive a substrate toward a particular transition state. Finally, like lysozyme, many enzymes participate intimately in the reaction by briefly forming a covalent bond between the substrate and a side chain of the enzyme. Subsequent steps in the reaction restore the side chain to its original state, so that the enzyme remains unchanged after the reaction.

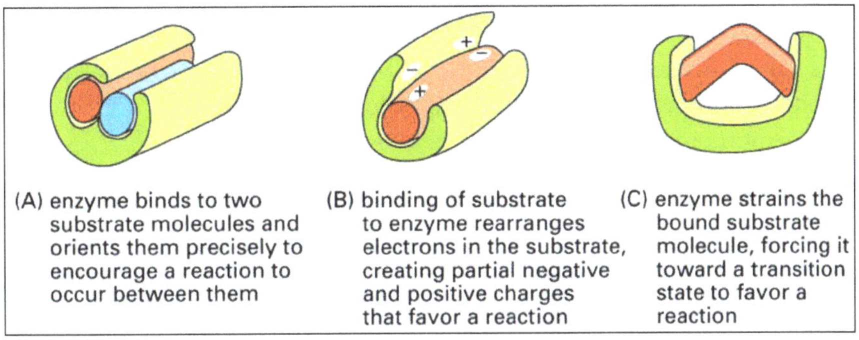

Figure: Some general strategies of enzyme catalysis.

Figure shows (A) Holding substrates together in a precise alignment. (B) Charge stabilization of reaction intermediates. (C) Altering bond angles in the substrate to increase the rate of a particular reaction.

Tightly Bound Small Molecules Add Extra Functions to Proteins

Although we have emphasized the versatility of proteins as chains of amino acids that perform different functions, there are many instances in which the amino acids by themselves are not enough. Just as humans employ tools to enhance and extend the capabilities of their hands, proteins often use small non-protein molecules to perform functions that would be difficult or impossible to do with amino acids alone. Thus, the signal receptor protein rhodopsin, which is made by the photoreceptor cells in the retina, detects light by means of a small molecule, retinal, embedded in the protein. Retinal changes its shape when it absorbs a photon of light, and this change causes the protein to trigger a cascade of enzymatic reactions that eventually leads to an electrical signal being carried to the brain.

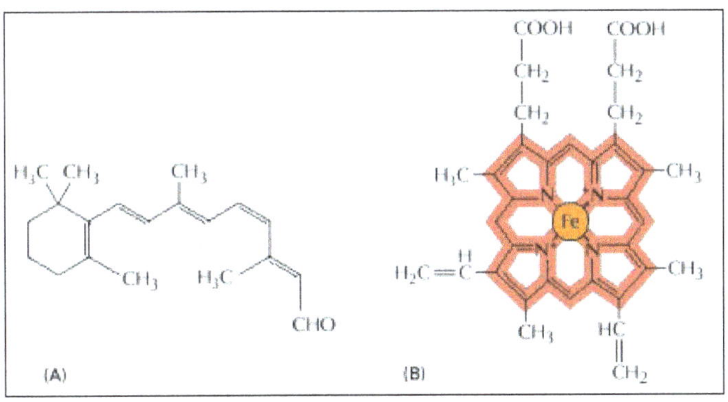

Figure: Retinal and heme.

Figure shows (A) The structure of retinal, the light-sensitive molecule attached to rhodopsin in the eye. (B) The structure of a heme group. The carbon-containing heme ring is red and the iron atom at its center is orange. A heme group is tightly bound to each of the four polypeptide chains in hemoglobin, the oxygen-carrying protein whose structure is shown.

Another example of a protein that contains a nonprotein portion is hemoglobin. A molecule of hemoglobin carries four heme groups, ring-shaped molecules each with a single central iron atom. Heme gives hemoglobin (and blood) its red color. By binding reversibly to oxygen gas through its iron atom, heme enables hemoglobin to pick up oxygen in the lungs and release it in the tissues.

Sometimes these small molecules are attached covalently and permanently to their protein, thereby becoming an integral part of the protein molecule itself. We see that proteins are often anchored to cell membranes through covalently attached lipid molecules. And membrane proteins exposed on the surface of the cell, as well as proteins secreted outside the cell, are often modified by the covalent addition of sugars and oligosaccharides.

Enzymes frequently have a small molecule or metal atom tightly associated with their active site that assists with their catalytic function. Carboxypeptidase, for example, an enzyme that cuts polypeptide chains, carries a tightly bound zinc ion in its active site. During the cleavage of a peptide bond by carboxypeptidase, the zinc ion forms a transient bond with one of the substrate atoms, thereby assisting the hydrolysis reaction. In other enzymes, a small organic molecule serves a similar purpose. Such organic molecules are often referred to as coenzymes. An example is biotin, which is found in enzymes that transfer a carboxylate group (–COO–) from one molecule to another. Biotin participates in these reactions by forming a transient covalent bond to the –COO– group to be transferred, being better suited to this function than any of the amino acids used to make proteins.

Table: Many Vitamins Provide Critical Coenzymes for Human Cells.

VITAMIN	COENZYME	ENZYME-CATALYZED REACTIONS REQUIRING THESE COENZYMES
Thiamine (vitamin B_1)	Thiamine pyrophosphate	Activation and transfer of aldehydes
Riboflavin (vitamin B_2)	FADH	Oxidation–reduction
Niacin	NADH, NADPH	Oxidation–reduction
Pantothenic acid	Coenzyme A	Acyl group activation and transfer
Pyridoxine	Pyridoxal phosphate	Reactions involving amino acid activation
Biotin	Biotin	CO_2 activation and transfer
Lipoic acid	Lipoamide	Acyl group activation; oxidation–reduction
Folic acid	Tetrahydrofolate	Activation and transfer of single carbon groups
Vitamin B_{12}	Cobalamin coenzymes	Isomerization and methyl group transfers

Multienzyme Complexes Help to Increase the Rate of Cell Metabolism

The efficiency of enzymes in accelerating chemical reactions is crucial to the maintenance of life. Cells, in effect, must race against the unavoidable processes of decay, which—if left unattended—cause macromolecules to run downhill toward greater and greater disorder. If the rates of desirable reactions were not greater than the rates of competing side reactions, a cell would soon die. Some idea of the rate at which cell metabolism proceeds can be obtained by measuring the rate of ATP utilization. A typical mammalian cell "turns over" (i.e., hydrolyzes and restores by phosphorylation) its entire ATP pool once every 1 or 2 minutes. For each cell this turnover represents the utilization of roughly 107 molecules of ATP per second (or, for the human body, about 1 gram of ATP every minute).

The rates of reactions in cells are rapid because of the effectiveness of enzyme catalysis. Many important enzymes have become so efficient that there is no possibility of further useful improvement. The factor that limits the reaction rate is no longer the enzyme's intrinsic speed of action; rather, it is the frequency with which the enzyme collides with its substrate. Such a reaction is said to be diffusion-limited.

If an enzyme-catalyzed reaction is diffusion-limited, its rate depends on the concentration of both the enzyme and its substrate. If a sequence of reactions is to occur extremely rapidly, each metabolic intermediate and enzyme involved must be present in high concentration. However, given the enormous number of different reactions performed by a cell, there are limits to the concentrations of substrates that can be achieved. In fact, most metabolites are present in micro-molar (10–6 M) concentrations, and most enzyme concentrations are much lower. How is it possible, therefore, to maintain very fast metabolic rates?

The answer lies in the spatial organization of cell components. Reaction rates can be increased without raising substrate concentrations by bringing the various enzymes involved in a reaction sequence together to form a large protein assembly known as a multi-enzyme complex. Because this allows the product of enzyme A to be passed directly to enzyme B, and so on, diffusion rates need not be limiting, even when the concentrations of the substrates in the cell as a whole are very low. It is perhaps not surprising, therefore, that such enzyme complexes are very common, and they are involved in nearly all aspects of metabolism—including the central genetic processes of DNA, RNA, and protein synthesis. In fact, few enzymes in eucaryotic cells may be left to diffuse freely in solution; instead, most seem to have evolved binding sites that concentrate them with other proteins of related function in particular regions of the cell, thereby increasing the rate and efficiency of the reactions that they catalyze.

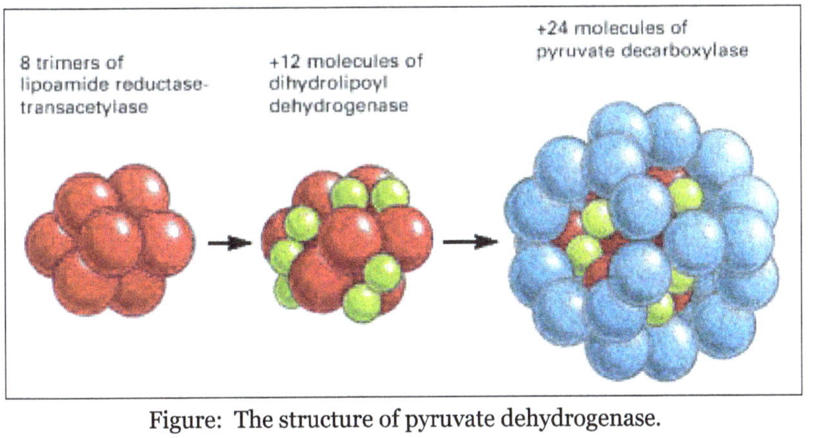

Figure: The structure of pyruvate dehydrogenase.

This enzyme complex catalyzes the conversion of pyruvate to acetyl CoA, as part of the pathway that oxidizes sugars to CO_2 and H_2O. It is an example of a large multienzyme complex in which reaction intermediates are passed directly from one enzyme to another.

Eucaryotic cells have yet another way of increasing the rate of metabolic reactions, using their intracellular membrane systems. These membranes can segregate particular substrates and the enzymes that act on them into the same membrane-enclosed compartment, such as the endoplasmic reticulum or the cell nucleus. If, for example, a compartment occupies a total of 10% of the volume of the cell, the concentration of reactants in the compartment may be increased as much as 10 times compared with the same cell with no compartmentalization. Reactions that would otherwise be limited by the speed of diffusion can thereby be speeded up by a factor of 10.

The Catalytic Activities of Enzymes are Regulated

A living cell contains thousands of enzymes, many of which operate at the same time and in the same small volume of the cytosol. By their catalytic action, these enzymes generate a complex web of metabolic pathways, each composed of chains of chemical reactions in which the product of one enzyme becomes the substrate of the next. In this maze of pathways, there are many branch points where different enzymes compete for the same substrate. The system is so complex that elaborate controls are required to regulate when and how rapidly each reaction occurs.

Regulation occurs at many levels. At one level, the cell controls how many molecules of each enzyme it makes by regulating the expression of the gene that encodes that enzyme. The cell also controls enzymatic activities by confining sets of enzymes to particular subcellular compartments, enclosed by distinct membranes. The rate of protein destruction by targeted proteolysis represents yet another important regulatory mechanism. But the most rapid and general process that adjusts reaction rates operates through a direct, reversible change in the activity of an enzyme in response to specific molecules that it encounters.

The most common type of control occurs when a molecule other than one of the substrates binds to an enzyme at a special regulatory site outside the active site, thereby altering the rate at which the enzyme converts its substrates to products. In feedback inhibition, enzyme acting early in a reaction pathway is inhibited by a late product of that pathway. Thus, whenever large quantities of the final product begin to accumulate, this product binds to the first enzyme and slows down its catalytic action, thereby limiting the further entry of substrates into that reaction pathway. Where pathways branch or intersect, there are usually multiple points of control by different final products, each of which works to regulate its own synthesis. Feedback inhibition can work almost instantaneously and is rapidly reversed when the level of the product falls.

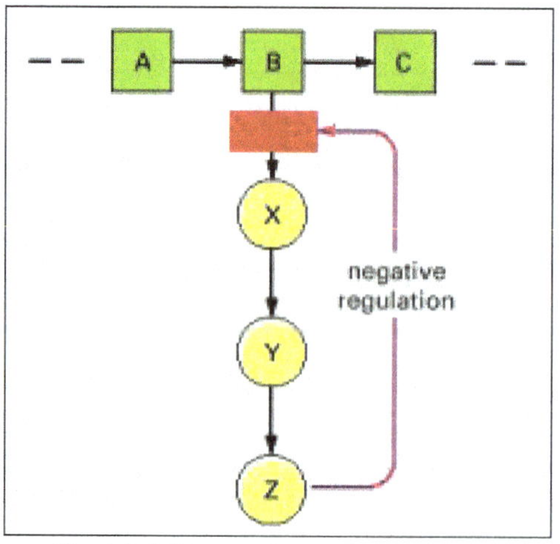

Figure: Feedback inhibition of a single biosynthetic pathway.

The end-product Z inhibits the first enzyme that is unique to its synthesis and thereby controls its own level in the cell. This is an example of negative regulation.

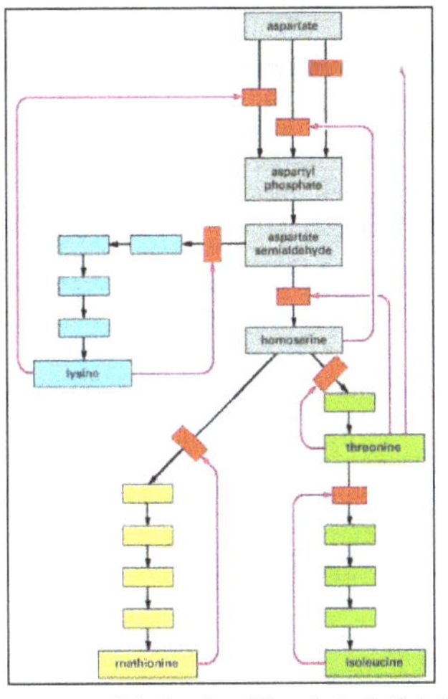

Figure: Multiple feedback inhibition.

In this example, which shows the biosynthetic pathways for four different amino acids in bacteria, the red arrows indicate positions at which products feedback to inhibit enzymes. Each amino acid controls the first enzyme specific to its own synthesis, thereby controlling its own levels and avoiding a wasteful buildup of intermediates. The products can also separately inhibit the initial set of reactions common to all the syntheses, in this case, three different enzymes catalyze the initial reaction, each inhibited by a different product.

Feedback inhibition is negative regulation: it prevents an enzyme from acting. Enzymes can also be subject to positive regulation, in which the enzyme's activity is stimulated by a regulatory molecule rather than being shut down. Positive regulation occurs when a product in one branch of the metabolic maze stimulates the activity of an enzyme in another pathway. As one example, the accumulation of ADP activates several enzymes involved in the oxidation of sugar molecules, thereby stimulating the cell to convert more ADP to ATP.

Allosteric Enzymes Have Two or more Binding Sites that Interact

One feature of feedback inhibition was initially puzzling to those who discovered it: The regulatory molecule often has a shape totally different from the shape of the substrate of the enzyme. This is why this form of regulation is termed allostery (from the Greek words allos, meaning "other," and stereos, meaning "solid" or "three-dimensional"). As more was learned about feedback inhibition, it was recognized that many enzymes must have at least two different binding sites on their surface—an active site that recognizes the substrates, and a regulatory site that recognizes a regulatory molecule. These two sites must somehow communicate in a way that allows the catalytic events at the active site to be influenced by the binding of the regulatory molecule at its separate site on the protein's surface.

The interaction between separated sites on a protein molecule is now known to depend on a conformational change in the protein: Binding at one of the sites causes a shift from one folded shape to a slightly different folded shape. During feedback inhibition, for example, the binding of an inhibitor at one site on the protein causes the protein to shift to a conformation in which its active site—located elsewhere in the protein—becomes incapacitated.

It is thought that most protein molecules are allosteric. They can adopt two or more slightly different conformations, and a shift from one to another caused by the binding of a ligand can alter their activity. This is true not only for enzymes but also for many other proteins—including receptors, structural proteins, and motor proteins. There is nothing mysterious about the allosteric regulation of these proteins: each conformation of the protein has somewhat different surface contours, and the protein's binding sites for ligands are altered when the protein changes shape. Each ligand stabilizes the conformation that it binds to most strongly, and thus—at high enough concentrations—tends to "switch" the protein toward the conformation it prefers.

Two Ligands whose Binding Sites are Coupled must Reciprocally Affect Each other's Binding

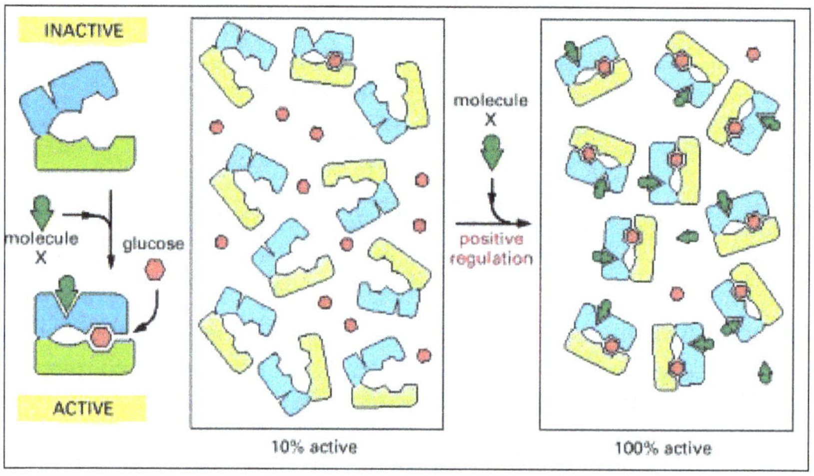

Figure: Positive regulation caused by conformational coupling between two distant binding sites.

The effects of ligand binding on a protein follow from a fundamental chemical principle known as linkage. Suppose, for example, that a protein that binds glucose also binds another molecule, X, at a distant site on the protein's surface. If the binding site for X changes shape as part of the conformational change induced by glucose binding, the binding sites for X and for glucose are said to be coupled. Whenever two ligands prefer to bind to the same conformation of an allosteric protein, it follows from basic thermodynamic principles that each ligand must increase the affinity of the protein for the other. Thus, if the shift of the protein in Figure below to the closed conformation that binds glucose best also causes the binding site for X to fit X better, then the protein will bind glucose more tightly when X is present than when X is absent.

In this example, both glucose and molecule X bind best to the closed conformation of a protein with two domains. Because both glucose and molecule X drive the protein toward its closed conformation, each ligand helps the other to bind. Glucose and molecule X are therefore said to bind cooperatively to the protein.

The relationships shown in above figure underlie all of cell biology. They seem so obvious in retrospect that we now take them for granted. But their discovery in the 1950s, followed by a general description of allostery in the early 1960s, was revolutionary at the time. Since molecule X in these examples binds at a site that is distinct from the site where catalysis occurs, it need have no chemical relationship to glucose or to any other ligand that binds at the active site. As we have just seen, for enzymes that are regulated in this way, molecule X can either turn the enzyme on (positive regulation) or turn it off (negative regulation). By such a mechanism, allosteric proteins serve as general switches that, in principle, allow one molecule in a cell to affect the fate of any other.

Symmetric Protein Assemblies Produce Cooperative Allosteric Transitions

A single subunit enzyme that is regulated by negative feedback can at most decrease from 90% to about 10% activity in response to a 100-fold increase in the concentration of an inhibitory ligand that it binds (Figure below, red line). Responses of this type are apparently not sharp enough for optimal cell regulation, and most enzymes that are turned on or off by ligand binding consist of symmetric assemblies of identical subunits. With this arrangement, the binding of a molecule of ligand to a single site on one subunit can trigger an allosteric change in the subunit that can be transmitted to the neighboring subunits, helping them to bind the same ligand. As a result, a cooperative allosteric transition occurs (Figure below, blue line), allowing a relatively small change in ligand concentration in the cell to switch the whole assembly from an almost fully active to an almost fully inactive conformation.

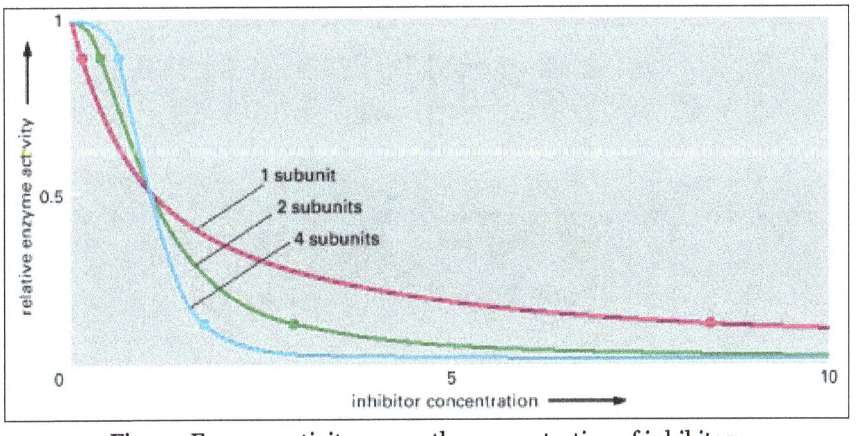

Figure: Enzyme activity versus the concentration of inhibitory ligand for single-subunit and multi-subunit allosteric enzymes.

For an enzyme with a single subunit (red line), a drop from 90% enzyme activity to 10% activity (indicated by the two dots on the curve) requires a 100-fold increase in the concentration of inhibitor. The enzyme activity is calculated from the simple equilibrium relationship $K = [I][P]/[IP]$, where P is active protein, I is inhibitor, and IP is the inactive protein bound to inhibitor. An identical curve applies to any simple binding interaction between two molecules, A and B. In contrast, a multisubunit allosteric enzyme can respond in a switchlike manner to a change in ligand concentration: the steep response is caused by a cooperative binding of the ligand molecules, as explained in Figure below. Here, the green line represents the idealized result expected for the cooperative binding of two inhibitory ligand molecules to an allosteric enzyme with two subunits, and the blue line shows the idealized response of an enzyme with four subunits. As indicated by the two dots on each of these curves, the more complex enzymes drop from 90% to 10% activity over a much narrower range of inhibitor concentration than does the enzyme composed of a single subunit.

The principles involved in a cooperative "all-or-none" transition are easiest to visualize for an enzyme that forms a symmetric dimer. In the example shown in Figure below, the first molecule of an inhibitory ligand binds with great difficulty since its binding destroys an energetically favorable interaction between the two identical monomers in the dimer. A second molecule of inhibitory ligand now binds more easily, however, because its binding restores the monomer–monomer contacts of a symmetric dimer (and also completely inactivates the enzyme).

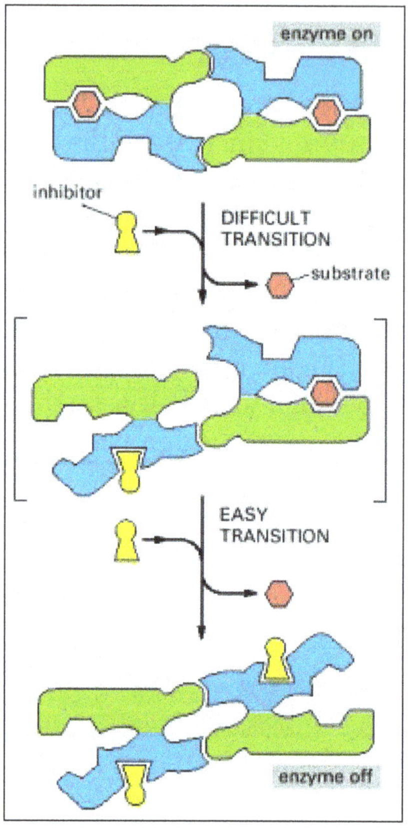

Figure: A cooperative allosteric transition in an enzyme composed of two identical subunits.

This diagram illustrates how the conformation of one subunit can influence that of its neighbor. The binding of a single molecule of an inhibitory ligand (yellow) to one subunit of the enzyme occurs with difficulty because it changes the conformation of this subunit and thereby destroys the symmetry of the enzyme. Once this conformational change has occurred, however, the energy gained by restoring the symmetric pairing interaction between the two subunits makes it especially easy for the second subunit to bind the inhibitory ligand and undergo the same conformational change. Because the binding of the first molecule of ligand increases the affinity with which the other subunit binds the same ligand, the response of the enzyme to changes in the concentration of the ligand is much steeper than the response of an enzyme with only one subunit. An even sharper response to a ligand can be obtained with larger assemblies, such as the enzyme formed from 12 polypeptide chains.

The Allosteric Transition in Aspartate Transcarbamoylase

One enzyme used in the early studies of allosteric regulation was aspartate transcarbamoylase from E. coli. It catalyzes the important reaction that begins the synthesis of the pyrimidine ring of C, U, and T nucleotides: carbamoylphosphate + aspartate → N-carbamoylaspartate. One of the final products of this pathway, cytosine triphosphate (CTP), binds to the enzyme to turn it off whenever CTP is plentiful.

Aspartate transcarbamoylase is a large complex of six regulatory and six catalytic subunits. The catalytic subunits are present as two trimers, each arranged like an equilateral triangle; the two trimers face each other and are held together by three regulatory dimers that form a bridge between them. The entire molecule is poised to undergo a concerted, all-or-none, allosteric transition between two conformations, designated as T (tense) and R (relaxed) states.

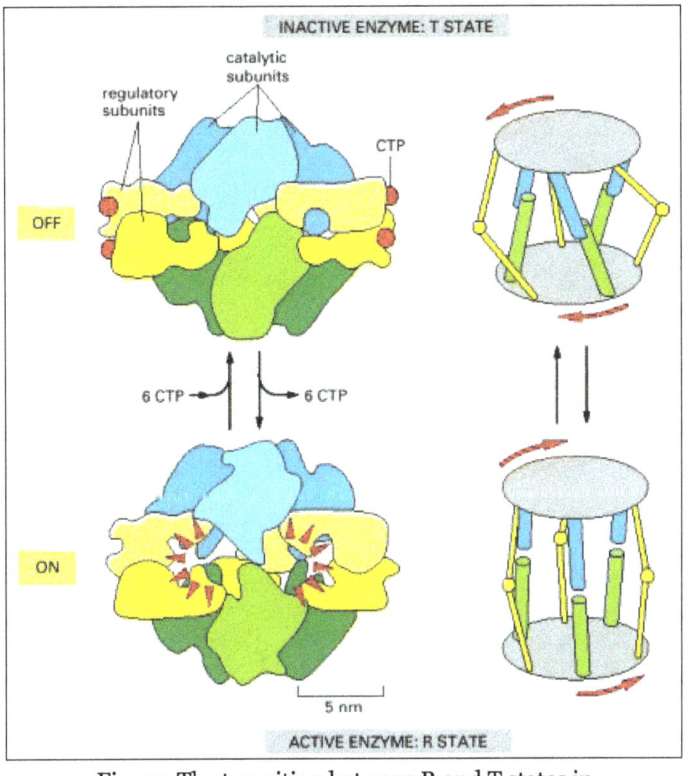

Figure: The transition between R and T states in the enzyme aspartate transcarbamoylase.

The enzyme consists of a complex of six catalytic subunits and six regulatory subunits, and the structures of its inactive (T state) and active (R state) forms have been determined by x-ray crystallography. The enzyme is turned off by feedback inhibition when CTP concentrations rise. Each regulatory subunit can bind one molecule of CTP, which is one of the final products in the pathway. By means of this negative feedback regulation, the pathway is prevented from producing more CTP than the cell needs.

The binding of substrates (carbamoylphosphate and aspartate) to the catalytic trimers drives aspartate transcarbamoylase into its catalytically active R state, from which the regulatory CTP molecules dissociate. By contrast, the binding of CTP to the regulatory dimers converts the enzyme to the inactive T state, from which the substrates dissociate. This tug-of-war between CTP and substrates is identical in principle to that described previously in above Figure for a simpler allosteric protein. But because the tug-of-war occurs in a symmetric molecule with multiple binding sites, the enzyme undergoes a cooperative allosteric transition that will turn it on suddenly as substrates accumulate (forming the R state) or shut it off rapidly when CTP accumulates (forming the T state).

A combination of biochemistry and x-ray crystallography has revealed many fascinating details of this allosteric transition. Each regulatory subunit has two domains, and the binding of CTP causes the two domains to move relative to each other, so that they function like a lever that rotates the two catalytic trimers and pulls them closer together into the T state. When this occurs, hydrogen bonds form between opposing catalytic subunits that help widen the cleft that forms the active site within each catalytic subunit, thereby destroying the binding sites for the substrates. Adding large amounts of substrate has the opposite effect, favoring the R state by binding in the cleft of each catalytic subunit and opposing the above conformational change. Conformations that are intermediate between R and T are unstable, so that the enzyme mostly clicks back and forth between its R and T forms, producing a mixture of these two species in proportions that depend on the relative concentrations of CTP and substrates.

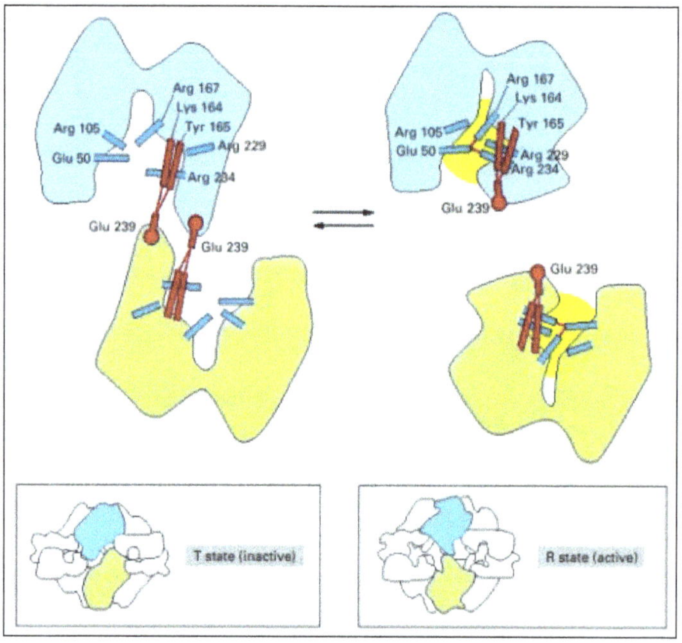

Figure: Part of the on–off switch in the catalytic subunits of aspartate transcarbamoylase.

Changes in the indicated hydrogen-bonding interactions are partly responsible for switching this enzyme's active site between active (yellow) and inactive conformations. Hydrogen bonds are indicated by thin red lines. The amino acids involved in the subunit–subunit interaction are shown in red, while those that form the active site of the enzyme are shown in blue. The large drawings show the catalytic site in the interior of the enzyme; the boxed sketches show the same subunits viewed from the enzyme's external surface.

Many Changes in Proteins are Driven by Phosphorylation

Enzymes are regulated by more than the binding of small molecules. A second method that is commonly used by eucaryotic cells to regulate a protein's function is the covalent addition of a phosphate group to one of its amino acid side chains. Such phosphorylation events can affect the protein in two important ways.

First, because each phosphate group carries two negative charges, the enzyme-catalyzed addition of a phosphate group to a protein can cause a major conformational change in the protein by, for example, attracting a cluster of positively charged amino acid side chains. This can, in turn, affect the binding of ligands elsewhere on the protein surface, dramatically changing the protein's activity through an allosteric effect. Removal of the phosphate group by a second enzyme returns the protein to its original conformation and restores its initial activity.

Second, an attached phosphate group can form part of a structure that is directly recognized by binding sites of other proteins. As previously discussed, certain small protein domains, called modules, appear very frequently in larger proteins. A large number of these modules provide binding sites for attaching their protein to phosphorylated peptides in other protein molecules. Several other types of modules recognize phosphorylated serine or threonine side chains in a specific context. As a result, protein phosphorylation and dephosphorylation events have a major role in driving the regulated assembly and disassembly of protein complexes.

Through such effects, reversible protein phosphorylation controls the activity structure and cellular localization of many types of proteins in eucaryotic cells. In fact, this regulation is so extensive that more than one-third of the 10,000 or so proteins in a typical mammalian cell are thought to be phosphorylated at any given time—many with more than one phosphate. As might be expected, the addition and removal of phosphate groups from specific proteins often occur in response to signals that specify some change in a cell's state. For example, the complicated series of events that takes place as a eucaryotic cell divides is largely timed in this way, and many of the signals mediating cell–cell interactions are relayed from the plasma membrane to the nucleus by a cascade of protein phosphorylation events.

A Eucaryotic Cell Contains a Large Collection of Protein Kinases and Protein Phosphatases

Protein phosphorylation involves the enzyme-catalyzed transfer of the terminal phosphate group of an ATP molecule to the hydroxyl group on a serine, threonine, or tyrosine side chain of the protein. This reaction is catalyzed by a protein kinase, and the reaction is essentially unidirectional because of the large amount of free energy released when the phosphate–phosphate bond in ATP is broken to produce ADP. The reverse reaction of phosphate removal, or dephosphorylation, is instead catalyzed by a protein phosphatase. Cells contain hundreds of different protein kinases, each responsible for phosphorylating a different protein or set of proteins. There are also many different protein phosphatases; some of these are highly specific and remove phosphate groups from only one or a few proteins, whereas others act on a broad range of proteins and are targeted to specific substrates by regulatory subunits. The state of phosphorylation of a protein at any moment, and thus its activity, depends on the relative activities of the protein kinases and phosphatases that modify it.

The protein kinases that phosphorylate proteins in eucaryotic cells belong to a very large family of enzymes, which share a catalytic (kinase) sequence of 250 amino acids. The various family members contain different amino acid sequences on either side of the kinase sequence, and often have short amino acid sequences inserted into loops within it (red arrowheads in Figure below). Some of these additional amino acid sequences enable each kinase to recognize the specific set of proteins it phosphorylates, or to bind to structures that localize it in specific regions of the cell. Other parts of the protein allow the activity of each enzyme to be tightly regulated, so it can be turned on and off in response to different specific signals.

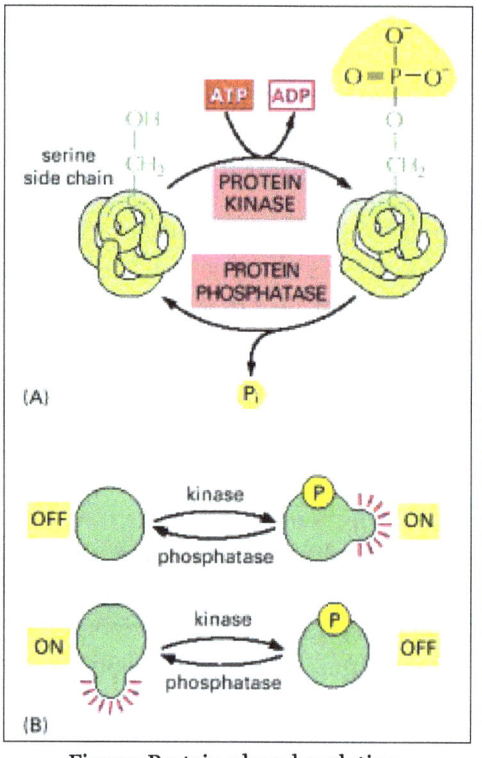

Figure: Protein phosphorylation.

Many thousands of proteins in a typical eucaryotic cell are modified by the covalent addition of a phosphate group. (A) The general reaction, shown here, entails the transfer of a phosphate group from ATP to an amino acid side chain of the target protein by a protein kinase. Removal of the phosphate group is catalyzed by a second enzyme, a protein phosphatase. In this example, the phosphate is added to a serine side chain; in other cases, the phosphate is instead linked to the –OH group of a threonine or a tyrosine in the protein. (B) The phosphorylation of a protein by a protein kinase can either increase or decrease the protein's activity, depending on the site of phosphorylation and the structure of the protein.

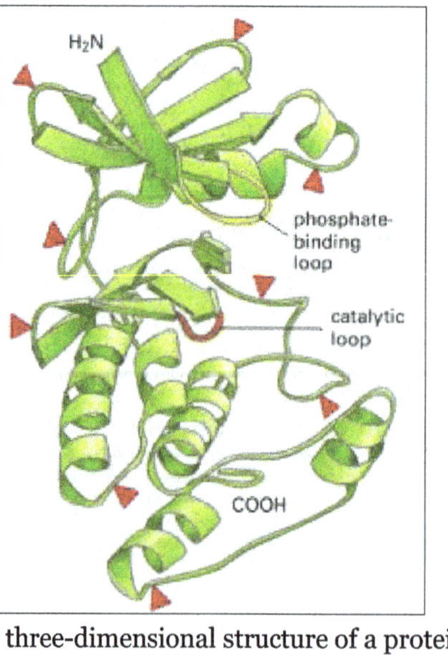

Figure: The three-dimensional structure of a protein kinase.

Superimposed on this structure are red arrowheads to indicate sites where insertions of 5–100 amino acids are found in some members of the protein kinase family. These insertions are located in loops on the surface of the enzyme where other ligands interact with the protein. Thus, they distinguish different kinases and confer on them distinctive interactions with other proteins. The ATP (which donates a phosphate group) and the peptide to be phosphorylated are held in the active site, which extends between the phosphate-binding loop (yellow) and the catalytic loop (orange).

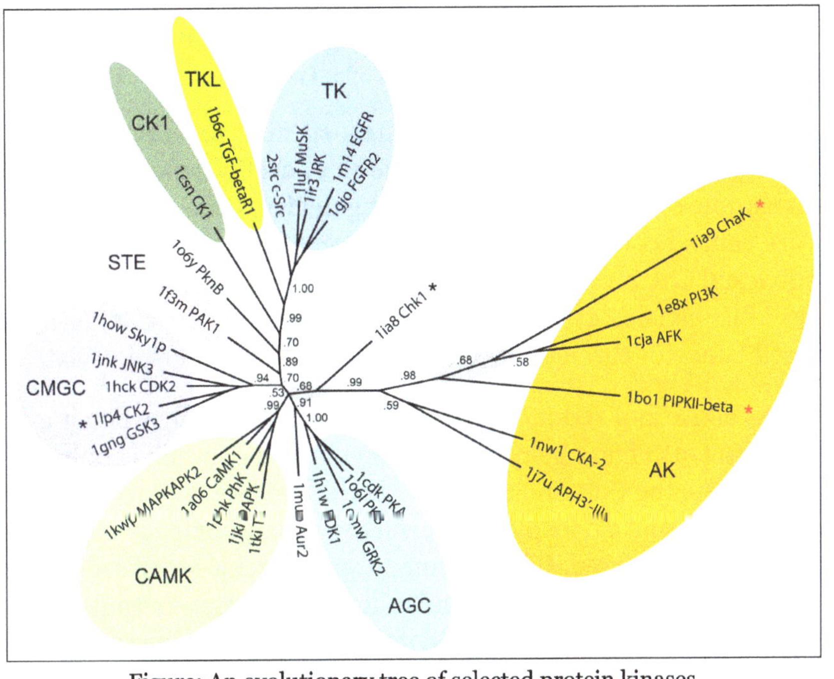

Figure: An evolutionary tree of selected protein kinases.

By comparing the number of amino acid sequence differences between the various members of a protein family, one can construct an "evolutionary tree" that is thought to reflect the pattern of gene duplication and divergence that gave rise to the family. An evolutionary tree of protein kinases is shown in Figure below. Not surprisingly, kinases with related functions are often located on nearby branches of the tree: the protein kinases involved in cell signaling that phosphorylate tyrosine side chains, for example, are all clustered in the top left corner of the tree. The other kinases shown phosphorylate either a serine or a threonine side chain, and many are organized into clusters that seem to reflect their function—in trans-membrane signal transduction, intracellular signal amplification, cell-cycle control, and so on. Although a higher eucaryotic cell contains hundreds of such enzymes, and the human genome codes for more than 500.

As a result of the combined activities of protein kinases and protein phosphatases, the phosphate groups on proteins are continually turning over—being added and then rapidly removed. Such phosphorylation cycles may seem wasteful, but they are important in allowing the phosphorylated proteins to switch rapidly from one state to another: the more rapid the cycle, the faster the state of phosphorylation of a population of protein molecules can change in response to a sudden stimulus that changes the phosphorylation rate. The energy required to drive the phosphorylation cycle is derived from the free energy of ATP hydrolysis, one molecule of which is consumed for each phosphorylation event.

The Regulation of Cdk and Src Protein Kinases Shows how a Protein can Function as a Microchip

The hundreds of different protein kinases in a eucaryotic cell are organized into complex networks of signaling pathways that help to coordinate the cell's activities, drive the cell cycle, and relay signals into the cell from the cell's environment. Many of the extracellular signals involved need to be both integrated and amplified by the cell. Individual protein kinases (and other signaling proteins) serve as input–output devices, or "microchips," in the integration process. An important part of the input to these proteins comes from the control that is exerted by phosphates added and removed from them by protein kinases and protein phosphatases, respectively, in the signaling network.

For a protein that is phosphorylated at multiple sites, specific sets of phosphate groups serve to activate the protein, while other sets can inactivate it. A cyclin-dependent protein kinase (Cdk) represents a good example of such a signal processing device. Kinases in this class phosphorylate serines, and they are central components of the cell-cycle control system in eucaryotic cells. In a vertebrate cell, individual Cdk enzymes turn on and off in succession, as a cell proceeds through the different phases of its division cycle. When a particular one of these kinases is on, it influences various aspects of cell behavior through effects on the proteins it phosphorylates.

A Cdk protein becomes active as a serine/threonine protein kinase only when it is bound to a second protein called a cyclin. But, as Figure below shows, the binding of cyclin is only one of three distinct "inputs" required to activate the Cdk. In addition to cyclin binding, a phosphate must be added to a specific threonine side chain, and a phosphate elsewhere in the protein (covalently bound to a specific tyrosine side chain) must be removed. Cdk thus monitors a specific set of cell components—a cyclin, a protein kinase, and a protein phosphatase—and it acts as an input–output device that turns on if, and only if, each of these components has attained its appropriate activity state. Some cyclins rise and fall in concentration in step with the cell cycle, increasing gradually in amount until they are suddenly destroyed at a particular point in the cycle. The sudden destruction of a cyclin (by targeted proteolysis) immediately shuts off its partner Cdk enzyme, and this triggers a specific step in the cell cycle.

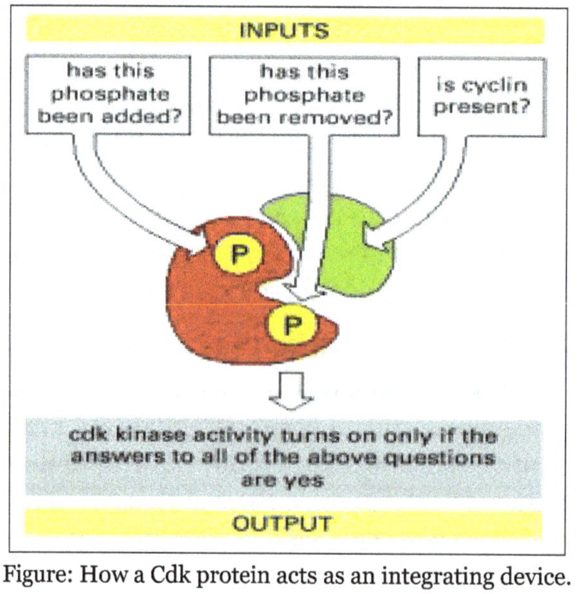

Figure: How a Cdk protein acts as an integrating device.

A similar type of microchip behavior is exhibited by the Src family of protein kinases. The Src protein (pronounced "sarc") was the first tyrosine kinase to be discovered, and it is now known to be part of a subfamily of nine very similar protein kinases, which are found only in multicellular animals. As indicated by the evolutionary tree in Figure, sequence comparisons suggest that tyrosine kinases as a group were a relatively late innovation that branched off from the serine/threonine kinases, with the Src subfamily being only one subgroup of the tyrosine kinases created in this way.

The Src protein and its homologs contain a short N-terminal region that becomes covalently linked to a strongly hydrophobic fatty acid, which holds the kinase at the cytoplasmic face of the plasma membrane. Next come two peptide-binding modules, a Src homology 3 (SH3) domain and a SH2 domain, followed by the kinase catalytic domains. These kinases normally exist in an inactive conformation, in which a phosphorylated tyrosine near the C-terminus is bound to the SH2 domain, and the SH3 domain is bound to an internal peptide in a way that distorts the active site of the enzyme and helps to render it inactive.

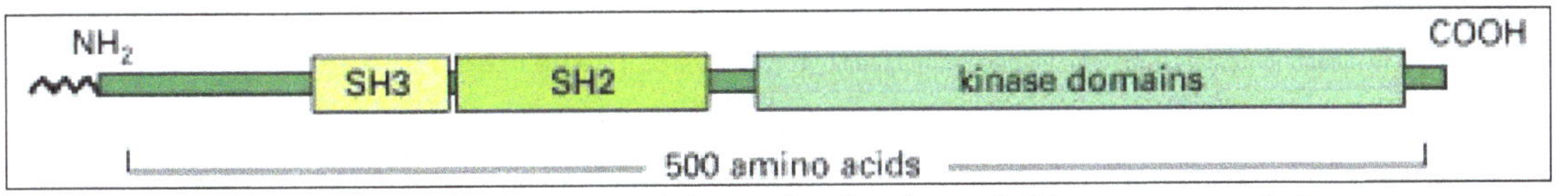

Figure: The domain structure of the Src family of protein kinases, mapped along the amino acid sequence.

Turning the kinase on involves at least two specific inputs: removal of the C-terminal phosphate and the binding of the SH3 domain by a specific activating protein. As for the Cdk protein, the activation of the Src kinase signals that a particular set of separate upstream events has been completed. Thus, both the Cdk and Src families of proteins serve as specific signal integrators, helping to generate the complex web of information-processing events that enable the cell to compute logical responses to a complex set of conditions.

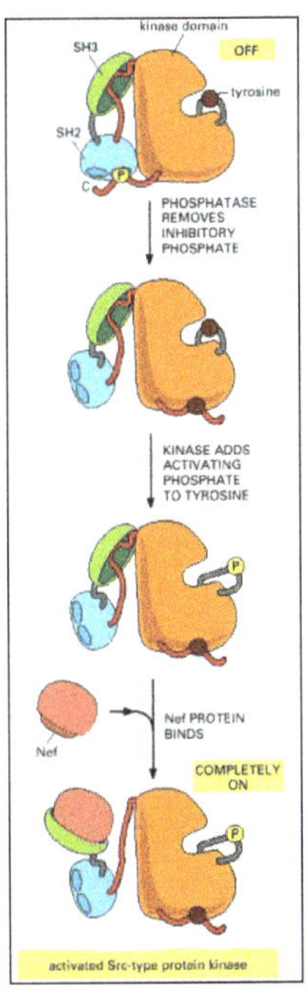

Figure: The activation of a Src-type protein kinase by two sequential events.

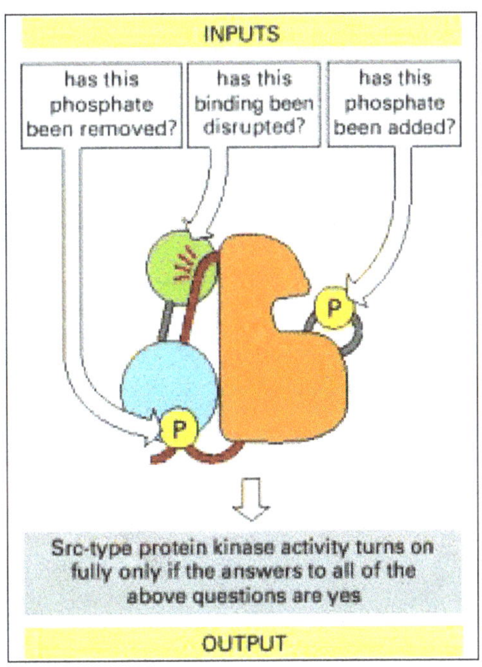

Figure: How a Src-type protein kinase acts as an integrating device.

The disruption of the SH3 domain interaction (green) can involve replacing its binding to the indicated red linker region with a tighter interaction with the Nef protein.

Proteins that Bind and Hydrolyze GTP are Ubiquitous Cellular Regulators

The addition or removal of phosphate groups on a protein can be used by a cell to control the protein's activity. In the examples discussed so far, the phosphate is transferred from an ATP molecule to an amino acid side chain of the protein in a reaction catalyzed by a specific protein kinase. Eucaryotic cells also have another way to control protein activity by phosphate addition and removal. In this case, the phosphate is not attached directly to the protein; instead, it is a part of the guanine nucleotide GTP, which binds very tightly to the protein. In general, proteins regulated in this way are in their active conformations with GTP bound. The loss of a phosphate group occurs when the bound GTP is hydrolyzed to GDP in a reaction catalyzed by the protein itself, and in its GDP bound state the protein is inactive. In this way, GTP-binding proteins act as on–off switches whose activity is determined by the presence or absence of an additional phosphate on a bound GDP molecule.

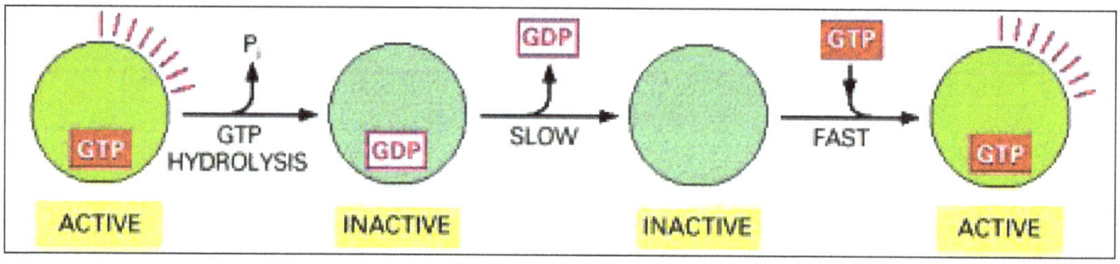

Figure: GTP-binding proteins as molecular switches.

The activity of a GTP-binding protein (also called a GTPase) generally requires the presence of a tightly bound GTP molecule (switch "on"). Hydrolysis of this GTP molecule produces GDP and inorganic phosphate (Pi), and it causes the protein to convert to a different, usually inactive, conformation (switch "off"). As shown here, resetting the switch requires the tightly bound GDP to dissociate, a slow step that is greatly accelerated by specific signals; once the GDP has dissociated, a molecule of GTP is quickly rebound.

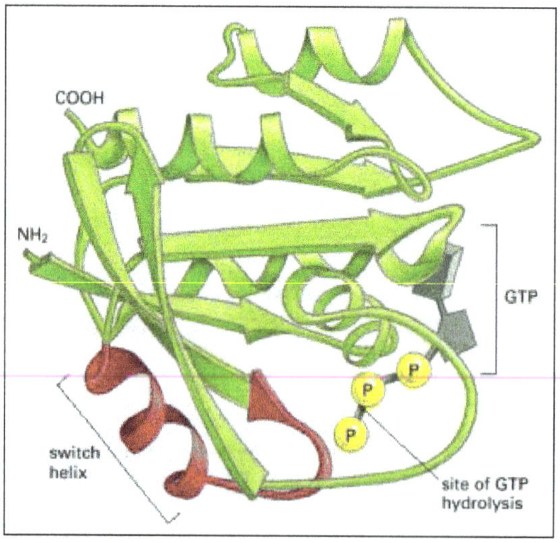

Figure: The structure of the Ras protein in its GTP-bound form.

This monomeric GTPase illustrates the structure of a GTP-binding domain, which is present in a large family of GTP-binding proteins. The red regions change their conformation when the GTP molecule is hydrolyzed to GDP and inorganic phosphate by the protein; the GDP remains bound to the protein, while the inorganic phosphate is released.

GTP-binding proteins (also called GTPases because of the GTP hydrolysis they catalyze) constitute a large family of proteins that all contain variations on the same GTP-binding globular domain. When the tightly bound GTP is hydrolyzed to GDP, this domain undergoes a conformational change that inactivates it. The three-dimensional structure of a prototypical member of this family, the monomeric GTPase called Ras, is shown in figure below.

The Ras protein has an important role in cell signaling. In its GTP-bound form, it is active and stimulates a cascade of protein phosphorylations in the cell. Most of the time, however, the protein is in its inactive, GDP-bound form. It becomes active when it exchanges its GDP for a GTP molecule in response to extracellular signals, such as growth factors, that bind to receptors in the plasma membrane.

Regulatory Proteins Control the Activity of Gtp-Binding Proteins by Determining Whether GTP or GDP is Bound

GTP-binding proteins are controlled by regulatory proteins that determine whether GTP or GDP is bound, just as phosphorylated proteins are turned on and off by protein kinases and protein phosphatases. Thus, Ras is inactivated by a GTPase-activating protein (GAP), which binds to the Ras protein and induces it to hydrolyze its bound GTP molecule to GDP—which remains tightly bound—and inorganic phosphate (Pi)—which is rapidly released. The Ras protein stays in its inactive, GDP-bound conformation until it encounters a guanine nucleotide exchange factor (GEF), which binds to GDP-Ras and causes it to release its GDP. Because the empty nucleotide-binding site is immediately filled by a GTP molecule (GTP is present in large excess over GDP in cells), the GEF activates Ras by indirectly adding back the phosphate removed by GTP hydrolysis. Thus, in a sense, the roles of GAP and GEF are analogous to those of a protein phosphatase and a protein kinase, respectively.

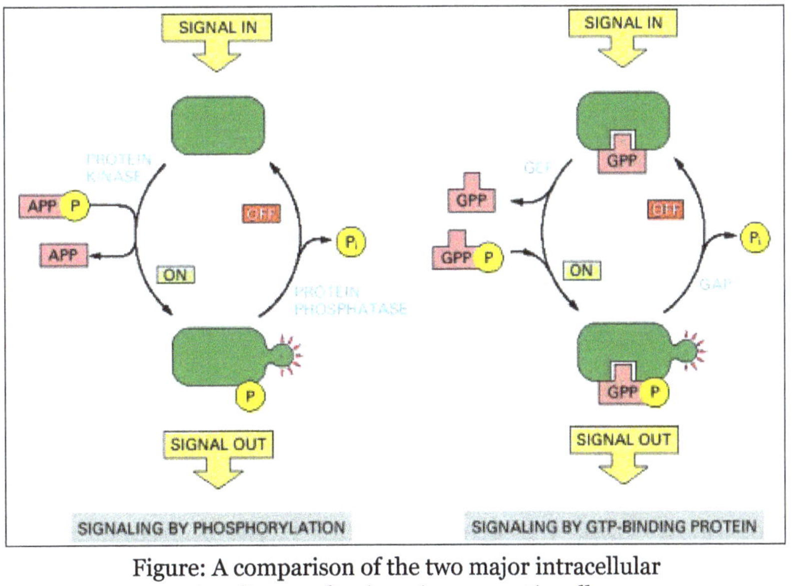

Figure: A comparison of the two major intracellular signaling mechanisms in eucaryotic cells.

In both cases, a signaling protein is activated by the addition of a phosphate group and inactivated by the removal of this phosphate. To emphasize the similarities in the two pathways, ATP and GTP are drawn as APPP and GPPP, and ADP and GDP as APP and GPP, respectively. As shown in Figure, the addition of a phosphate to a protein can also be inhibitory.

Large Protein Movements can be Generated from Small Ones

The Ras protein belongs to a large superfamily of monomeric GTPases, each of which consists of a single GTP-binding domain of about 200 amino acids. Over the course of evolution, this domain has also become joined to larger proteins with additional domains, creating a large family of GTP-binding proteins. Family members include the receptor-associated trimeric G proteins involved in cell signaling, proteins regulating the traffic of vesicles between intracellular compartments, and proteins that bind to transfer RNA and are required as assembly factors for protein synthesis on the ribosome. In each case, an important biological activity is controlled by a change in the protein's conformation that is caused by GTP hydrolysis in a Ras-like domain.

The EF-Tu protein provides a good example of how this family of proteins works. EF-Tu is an abundant molecule that serves as an elongation factor (hence the EF) in protein synthesis, loading each amino-acyl tRNA molecule onto the ribosome. The tRNA molecule forms a tight complex with the GTP-bound form of EF-Tu. In this complex, the amino acid attached to the tRNA is masked. Unmasking is required for the tRNA to transfer its amino acid in protein synthesis, and it occurs on the ribosome when the GTP bound to EF-Tu is hydrolyzed, allowing the tRNA to dissociate. Since the GTP hydrolysis is triggered by a proper fit of the tRNA to the mRNA molecule on the ribosome, the EF-Tu serves as an assembly factor that discriminates between correct and incorrect mRNA–tRNA pairings.

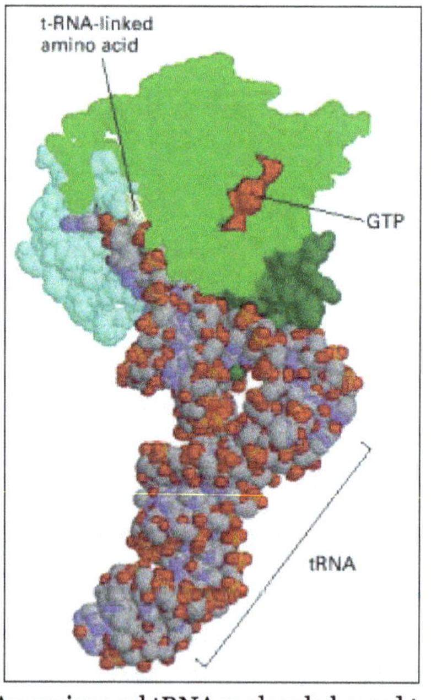

Figure: An aminoacyl tRNA molecule bound to EF-Tu.

The three domains of the EF-Tu protein are colored differently, to match figure. This is a bacterial protein; however, a very similar protein exists in eukaryotes, where it is called EF-1.

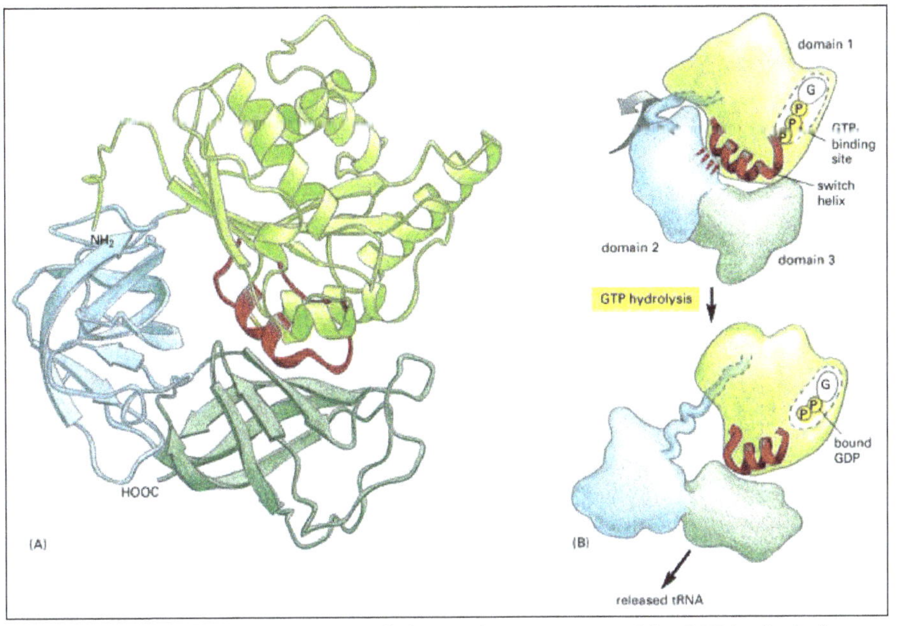

Figure: The large conformational change in EF-Tu caused by GTP hydrolysis.

Figure shows: (A) The three-dimensional structure of EF-Tu with GTP bound. The domain at the top is homologous to the Ras protein, and its red α helix is the switch helix, which moves after GTP hydrolysis. (B) The change in the conformation of the switch helix in domain 1 causes domains 2 and 3 to rotate as a single unit by about 90° toward the viewer, which releases the tRNA that was shown bound to this structure.

Comparison of the three-dimensional structure of EF-Tu in its GTP-bound and GDP-bound forms reveals how the unmasking of the tRNA occurs. The dissociation of the inorganic phosphate group (P_i), which follows the reaction GTP → GDP + P_i, causes a shift of a few tenths of a nanometer at the GTP-binding site, just as it does in the Ras protein. This tiny movement, equivalent to a few times the diameter of a hydrogen atom, causes a conformational change to propagate along a crucial piece of α helix, called the switch helix, in the Ras-like domain of the protein. The switch helix seems to serve as a latch that adheres to a specific site in another domain of the molecule, holding the protein in a "shut" conformation. The conformational change triggered by GTP hydrolysis causes the switch helix to detach, allowing separate domains of the protein to swing apart, through a distance of about 4 nm. This releases the bound tRNA molecule, allowing its attached amino acid to be used.

One can see from this example how cells have exploited a simple chemical change that occurs on the surface of a small protein domain to create a movement 50 times larger. Dramatic shape changes of this type also underlie the very large movements that occur in motor proteins.

Motor Proteins Produce Large Movements in Cells

These motor proteins generate the forces responsible for muscle contraction and the crawling and swimming of cells. Motor proteins also power smaller-scale intracellular movements: they help to move chromosomes to opposite ends of the cell during mitosis, to move organelles along molecular tracks within the cell, and to move enzymes along a DNA strand during the synthesis of a new DNA molecule. All these fundamental processes depend on proteins with moving parts that operate as force-generating machines.

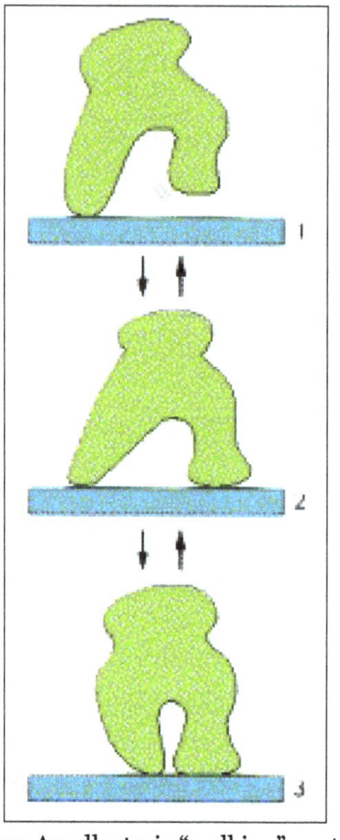

Figure: An allosteric "walking" protein.

How do these machines work? In other words, how are shape changes in proteins used to generate directed movements in cells? If, for example, a protein is required to walk along a narrow thread such as a DNA molecule, it can do this by undergoing a series of conformational changes. With nothing to drive these changes in an orderly sequence, however, they are perfectly reversible, and the protein can wander randomly back and forth along the thread. We can look at this situation in another way. Since the directional movement of a protein does work, the laws of thermodynamics demand that such movement utilize free energy from some other source (otherwise the protein could be used to make a perpetual motion machine). Therefore, without an input of energy, the protein molecule can only wander aimlessly.

Although its three different conformations allow it to wander randomly back and forth while bound to a thread or a filament, the protein cannot move uniformly in a single direction. How, then, can one make the series of conformational changes unidirectional? To force the entire cycle to proceed in one direction, it is enough to make any one of the changes in shape irreversible. For most proteins that are able to walk in one direction for long distances, this is achieved by coupling one of the conformational changes to the hydrolysis of an ATP molecule bound to the protein. The mechanism is similar to the one just discussed that drives allosteric protein shape changes by GTP hydrolysis. Because a great deal of free energy is released when ATP (or GTP) is hydrolyzed, it is very unlikely that the nucleotide-binding protein will undergo the reverse shape change needed for moving backward—since this would require that it also reverse the ATP hydrolysis by adding a phosphate molecule to ADP to form ATP.

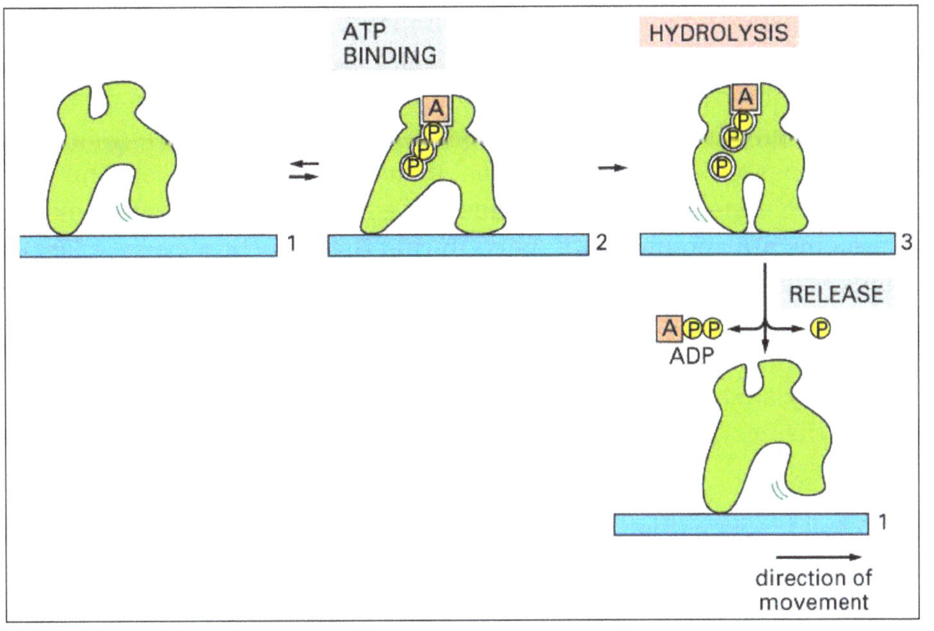

Figure: An allosteric motor protein.

In the model shown in Figure below, ATP binding shifts a motor protein from conformation 1 to conformation 2. The bound ATP is then hydrolyzed to produce ADP and inorganic phosphate (P_i), causing a change from conformation 2 to conformation 3. Finally, the release of the bound ADP and Pi drives the protein back to conformation 1. Because the transition 2 → 3 is driven by the energy provided by ATP hydrolysis, this series of conformational changes is effectively irreversible. Thus, the entire cycle goes in only one direction, causing the protein molecule to walk continuously to the right in this example.

The transition between three different conformations includes a step driven by the hydrolysis of a bound ATP molecule, and this makes the entire cycle essentially irreversible. By repeated cycles, the protein therefore moves continuously to the right along the thread.

Many motor proteins generate directional movement in this general way, including the muscle motor protein myosin, which walks along actin filaments to generate muscle contraction, and the kinesin proteins that walk along microtubules. These movements can be rapid: some of the motor proteins involved in DNA replication (the DNA helicases) propel themselves along a DNA strand at rates as high as 1000 nucleotides per second.

Membrane-bound Transporters Harness Energy to Pump Molecules through Membranes

We have thus far seen how allosteric proteins can act as microchips (Cdk and Src kinases), as assembly factors (EF-Tu), and as generators of mechanical force and motion (motor proteins). Allosteric proteins can also harness energy derived from ATP hydrolysis, ion gradients, or electron transport processes to pump specific ions or small molecules across a membrane.

One of the best understood pump proteins is the calcium transport protein from muscle cells. This protein, called the Ca^{2+} pump, is embedded in the membrane of a specialized organelle in muscle cells called the sarcoplasmic reticulum. The Ca^{2+} pump (also known as the Ca^{2+} ATPase) maintains the low cytoplasmic calcium concentration of resting muscle by pumping calcium out of the cytosol into the membrane-enclosed sarcoplasmic reticulum; then, in response to a nerve impulse, Ca^{2+} is rapidly released (through other channels) back into the cytosol to trigger muscle contraction. The Ca^{2+} pump is homologous to the Na^+-K^+ pump that maintains Na^+ and K^+ concentration differences across the plasma membrane, both being members of a family of P-type cation pumps—so named because these proteins are autophosphorylated during their reaction cycle. A large, cytoplasmic head region binds and hydrolyzes ATP, forming a covalent bond with the phosphate released. The protein thereby transiently shifts to a high-energy, phosphorylated state that is tightly bound to two Ca^{2+} ions that were picked up from the cytosol. This form of the protein then decays to a low-energy, phosphorylated state, which causes the Ca^{2+} ions to be released into the lumen. Dephosphorylation of the enzyme finally releases the phosphate and resets the protein for its next round of ion pumping.

The head region of all P-type cat-ion pumps is linked to a series of ten trans-membrane α helices, four of which form trans-membrane Ca^{2+} -binding sites for the Ca^{2+} pump. The three-dimensional structure of this protein has been deciphered by high-resolution electron microscopy and x-ray diffraction. This has enabled biologists to derive a molecular model for pump action based on extensive biochemical data on its normal and mutant forms.

As illustrated in Figure below, the trans-membrane α helices that bind the Ca^{2+} wind around each other and create a cavity for Ca^{2+} between the helices. ATP hydrolysis generates a series of major conformational changes in the cytoplasmic head, which—through its stalk connection to the trans-membrane helices—changes the structure and relative orientations of some of the helices in the membrane, thereby altering the cavity in a way that pushes the Ca^{2+} ions unidirectionally across the membrane. As for a motor protein, unidirectional transport of an ion requires the cycle to use energy, so as to impart a preferred directionality to the protein's conformational changes.

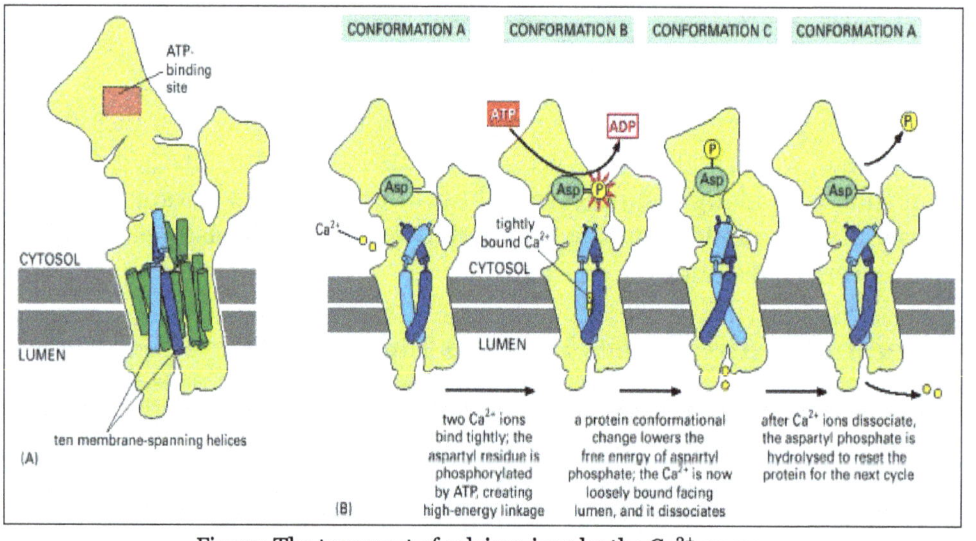

Figure: The transport of calcium ions by the Ca^{2+} pump.

(A) The structure of the pump protein, formed from a single subunit of 994 amino acids. (B) A model for ion pumping. For simplicity, only two of the four trans-membrane α helices that bind Ca^{2+} are shown.

Humans have invented many different types of mechanical pumps, and it should not be surprising that cells also contain membrane-bound pumps that function in other ways. Most notable are the rotary pumps that couple the hydrolysis of ATP to the transport of H^+ ions (protons). These pumps resemble miniature turbines, and they are used to acidify the interior of lysosomes and other eucaryotic organelles. Like other ion pumps that create ion gradients, they can function in reverse to catalyze the reaction ADP + P_i → ATP if there is a steep enough gradient across their membrane of the ion that they transport.

One such pump, the ATP synthase, harnesses a gradient of proton concentration produced by electron transport processes to produce most of the ATP used in the living world. This ubiquitous pump has such a central role in energy conversion.

Proteins Often Form Large Complexes that Function as Protein Machines

As one progresses from small, single-domain proteins to large proteins formed from many domains, the functions the proteins can perform become more elaborate. The most impressive tasks, however, are carried out by large protein assemblies formed from many protein molecules. Now that it is possible to reconstruct most biological processes in cell-free systems in the laboratory, it is clear that each of the central processes in a cell—such as DNA replication, protein synthesis, vesicle budding, or trans-membrane signaling—is catalyzed by a highly coordinated, linked set of ten or more proteins. In most such protein machines, the hydrolysis of bound nucleoside triphosphates (ATP or GTP) drives an ordered series of conformational changes in some of the individual protein subunits, enabling the ensemble of proteins to move coordinately. In this way, each of the appropriate enzymes can be placed directly into the positions where they are needed to perform successive reactions in a series. This is what occurs, for example, in protein synthesis on a ribosome—or in DNA replication, where a large multiprotein complex moves rapidly along the DNA.

Cells have evolved protein machines for the same reason that humans have invented mechanical and electronic machines. For accomplishing almost any task, manipulations that are spatially and temporally coordinated through linked processes are much more efficient than the sequential use of individual tools.

A Complex Network of Protein Interactions Underlies Cell Function

There are many challenges facing cell biologists in this "post-genome" era when complete genome sequences are known. One is the need to dissect and reconstruct each one of the thousands of protein machines that exist in an organism such as ourselves. To understand these remarkable protein complexes, each must be reconstituted from its purified protein parts—so that its detailed mode of operation can be studied under controlled conditions in a test tube, free from all other cell components. This alone is a massive task. But we now know that each of these subcomponents of a cell also interacts with other sets of macromolecules, creating a large network of protein–protein and protein–nucleic acid interactions throughout the cell. To understand the cell, therefore, one will need to analyze most of these other interactions as well.

Some idea of the complexity of intracellular protein networks can be gained from a particularly well-studied example: The many dozens of proteins that interact with the actin cytoskeleton in the yeast Saccharomyces cerevisiae.

The extent of such protein–Protein interactions can also be estimated more generally. In particular, large-scale efforts have been undertaken to detect these interactions using the two-hybrid screening method. Thus, for example, this method is being applied to a large set of the 6000 gene products produced by S. cerevisiae. As expected, the majority of the interactions that have been observed are between proteins in the same functional group. That is, proteins involved in cell-cycle control tend to interact extensively with each other, as do proteins involved with DNA synthesis, or DNA repair, and so on. But there are also a surprisingly large number of interactions between the protein members of different functional groups. These interactions are presumably important for coordinating cell functions, but most of them are not understood.

To produce this map, more than 1500 proteins were assigned to the indicated 31 functional groups. About 70 percent of the known protein–protein interactions were observed to occur between proteins assigned to the same functional group. However, as indicated, a surprisingly large number of interactions crossed between these groups. Each line in this diagram designates that more than 15 protein–protein interactions were observed between proteins in the two functional groups that are connected by that line. The three functional groups highlighted in yellow display at least 15 interactions with 10 or more of the 31 functional groups defined in the study.

References

- Protein-structure-14122136: nature.com, Retrieved 11, July 2020
- Protein, science: britannica.com, Retrieved 05, May 2020
- A-Protein-Structure, A-Understanding-Protein-Conformation, Secondary-Structure, Biochemistry: bio.libretexts.org, Retrieved 29, March 2020
- Proteins, biology, science, science-maths-technology: open.edu, Retrieved 11, August 2020
- Tertiary-structure: biologydictionary.net, Retrieved 17, July 2020

Chapter 2

Protein-Protein Interactions

Protein-protein interactions are physical contacts between proteins in the cell as a result of biochemical events. Hydrogen bonds, covalent bonds, non-covalent bonds, etc. are the various bonds which contribute in protein-protein interactions. This chapter touches on the mentioned topics in a detailed manner for the benefit of the reader.

Protein-Protein-Interactions (PPIs) are involved in almost all the cellular processes and understanding the structural basis of PPIs remains an important endeavor. The identification of the interface residues may shed light in many important aspects like drug development, elucidation of molecular pathways, generation of protein mimetics and understanding of disease mechanisms as well as development of docking methodologies to build structural models of protein complexes. Over the past few years, advances in high throughput PPI identification techniques, such as yeast two-hybrid analysis and affinity purification coupled with mass spectrometry, have enabled the researchers to identify sets of interacting proteins in yeast, Drosophila and other organisms. Unfortunately, these experimental methods do not provide residue level insight into the structure of the interactions between the proteins. The uses of X-Ray crystallography and nuclear magnetic resonance (NMR) spectroscopy to determine the structural basis of an interaction are time consuming and overall expensive. In response to these difficulties, a number of different bioinformatics algorithms with varying degrees of accuracies have been developed that use a wide variety of data sources to predict PPIs and modes of binding between proteins. Machine learning techniques such as Support Vector Machines (SVMs) and Random Forests (RFs) have been used recently to solve problems such as prediction of catalytic residues and prediction and analysis of structure-based PPI interfaces. Previous machine learning approaches to the PPI interface prediction problems used features pertaining to evolutionary amino acid sequence conservation, phylogeny, and GO (Gene Ontology) protein annotation and, in most of the cases, protein structures. Till date there are very few computational methods available that are based solely on protein sequences.

Most, if not all, of our biological functions rely on the activities of proteins, a versatile biomolecule. Proteins' functions depend on the interactions of it with its partner, which may be another protein or other biomolecules (like RNA or DNA or others), or an organic co-factor or inorganic prosthetic group (like Mg^{++}, or O_2, or Zn etc.). Thus the understandings of the interactions of protein with other partner proteins are very much essential to decipher the bio-molecular pathways leading to some physiological events. The protein-protein interactions (PPI) have a great role in many biological events like Enzyme substrate interactions or hormone-receptor binding or immunological functions (like antigen recognition and presentation etc.) to name a few of them, revealing the fact that an altered PPI can have a potential to disease onset. Thus exploring the molecular details of PPI will not only enrich our knowledge but also could be highly useful in modern medical biochemistry fields like molecular medicine research and other areas.

PPI are mediated by a varied set of proteins but often done by the proteins having a specific quaternary structure. These types of proteins are generally found in all cellular places like membranes (as transporters or channels or receptors etc.), cytosol (like ribosomes, carrier proteins, chaperones and several other enzymes) and in different sub cellular organelles. Energy transfer and harvesting, enzyme co-operativity, structural integrity (like cytoskeletons forming and deforming), cellular motion (the relative movement of ciliary or flagellar fibrils), functions of several molecular motors and switches and signal transduction pathways inside a cell have great impact from PPI of multimeric proteins. Decades of research shed light on this part to elucidate the PPI. A brief year-wise list of early works has been provided just to summarize some of them. Unintentional absence of other relevant and good work is regretted by the authors.

Proteins

Proteins are polymers, an assembly of its single constituent units, of L (-) amino acids, with S as the stereo-configuration. Amino acids have one carboxylic (-COOH) group, one amino ($-NH_2$) group, one hydrogen (-H) and one side chain (R), centering one chiral carbon or the alpha carbon (except in glycine, where the centre is achiral). Depending on the physico-chemical properties of the side chains, amino acids are grouped into twenty proteinergic amino acids. In a protein chain, amino acids form a highly stable covalent bond viz. peptide bond (-CO-NH-) between the previous residues carboxylic group to next amino group. Side chains of different amino acids are important for PPI, as they actively take part in it.

PPI

Proteins in biological system work more commonly in oligomeric form, rather than in isolation or in monomeric form. Oligomeric proteins are associations of two or more proteins. While an oligomer of two proteins is preferentially called a dimer. Each of the constituent protein chains in the oligomer is known as Protomer. Studies have revealed protein oligomerization has advantageous role on their stability, functions and accessibility and increased complexity. It also has been shown unwanted or altered interaction may potentially lead to pathogenic conditions. This functional improvement suggests an evolutionary gain in oligomerization. Thus interacting interface showed conserved amino acid residues at interacting surfaces (like hydrophobic protein core with polar surfaces). PPI, a very common physical process in cell, plays effective roles in many cellular physiology and metabolisms.

Enzyme regulation, stability and activation: Caspase-9 shows inactivity in monomeric form. Together with Apaf1, cofactor, and in the presence of Cytochrome C and ATP it forms an apoptosome, an oligomeric form that helps the enzyme to produce its active form by dimerization of self priming.

Local rise in concentration of biological macromolecules by decreasing degradation: Binding of 14-3-3 showed increased stability for Arylalkylamine N-Acetyl Transferase (AANAT), the penultimate enzyme for melatonin synthesis. And, thus the local concentration of the enzyme is increased in Pinealocyte and becomes the rate limiting step for the hormone biosynthesis.

Channels and receptor functions: Receptors like tyrosine kinase phosphates alpha present on the cell surface work mainly as a weak homodimer, as one monomer intrudes the active site of the other one. This prevents the unwanted transfer of molecules. Signal transmission. During acute inflammation a cytokine, Interleukin-12 signals to produce interferon-γ {IFN- ϒ}. Hetero-dimer of p35 and p40 subunits is the biologically active form of IL-12. Studies have shown though that p40 monomer has no effect on IL-12 functions, the p40 homo-dimer can bind to the IL-12 receptor, resulting in inhibition of IFN- ϒproduction.

Regulation of combinatorial specificity: Specific activity of enzymes like protein tyrosine phosphatases shows a regulation by target domains of substrate and catalytic domains of enzyme in a combinatorial fashion. Signaling protein molecules also show same type of combinatorial specificity during signal transmission. Allosteric regulation, activation and inhibition. During release and loading of Oxygen (O_2), Hemoglobin molecules show an allosteric regulation that gives rise to characteristic sigmoid curve of O_2 saturation. While Haemoglobins are unloaded with O_2, subunits of hemoglobin are in T (tense) state and after loading of O_2 it becomes to be in R (relaxed) state. Studies have shown T state has less affinity toward O_2 as the PPI between subunits constrained the structure to make the O_2 binding site available. Gradual binding of O_2 makes the subunit freer than T state, thus inducing R state and making the subunits less constrained to make the O_2 binding sites more available.

Hereditary information transfer: For instance, in Yeast a protein only mechanism is used to transfer phenotypic alterations to next generation. Self-perpetuating amyloidogenic conformations have been shown to act as hereditary material there. These are some of the well-known examples among many other functions of PPI.

Types of PPI

The PPIs can broadly be classified into several different categories. The categories include the type of interacting partners, stability of the PPI complexes, time-span of the interactions between the protein partners, nature of the interaface between the proteins.

Identity of Interacting Partners

If interacting protein chains are identical they form homooligomer, with structural symmetry. Non-identical partners form heterooligomers. Haemoglobin forms homo-tetramer and 14-3-3 beta isomer proteins form homodimer. G protein coupled receptors are example of hetero-oligomers.

Stability of Interacting Complexes

If protomers cannot exist in free form and only stable in multimeric association they form obligate oligomers (Homo-obligomers and/or heteroobligomers), like the Arc repressor dimer where dimerization is essential for DNA binding. When protomers can exist in free form as well, they are non-obligate partners like antigen-antibody complex or intracellular signal transmission complex. Mostly non obligate partners do not co-localize, thus in need of self-stability.

Lifetime of PPI

When an association between protomers is highly stable and need help from molecular switches to break them, they are permanent complexes. Hetero-trimeric G protein (Gα, Gβϒ and GDP) forms this type of PPI. In contrast to this, Sperm Lysin, a homo-dimer, and many other show a dynamic equilibrium of association and dissociation of oligo-meric form. This type of PPI is named as transient complexes (sometimes mentioned as Homo complex or hetero complex, depending on the partners).

Nature of the Interaction Interface

When protomers use the same interacting interface to join each other, they are isologous complex. Arc repressor and Lysine; show this type of PPI with a two-fold symmetry axis. They can only interact with other partners using a different interacting surface. In Heterologous assembly protomers use different interfaces to form PPI without any closed symmetry. Thus they have chances to aggregate infinitely. Amyloid aggregation might have this type of PPI, allowing them to form a long multimeric disposition. Also another type of grouping has been reported as Intra-domain, PPI within one structural domain of protein, and Domain-domain, PPI of different domains within a single chain or different chains of proteins.

Though this grouping could be helpful for further studies on PPI, sometimes protomers do not form any of these distinct types of PPIs. The characteristics of PPIs are regulated mainly by the cellular environment, physiology and the need of biological functions.

Details of PPI Interface

Oligomeric proteins do have selective advantages for benefits like surface area reduction, increased stability and novel functions via communication among subunits. The interaction surface of protomers is generally mentioned as interface. Some of the important characteristic properties of PPI interface are discussed.

A PPI interface has been defined at two different ways viz. at residue level and at atomic level. For both cases we have to learn about the concept of solvent accessible surface area (ASA), the extent to which a protein can come in contact with water. Based on relative asa values, ratio of actual asa of any residue in a protein chain compare to the maximal accessibility of each amino acid (ranges from 0%-100%), we can find out the interior residues (relative ASA ≤5%) and exterior residues (relative ASA >5%). If after dimerization any residue shows its' side chain ASA has been decreased by >1%, that residue is considered as interface residue. At atom level, if any atom shows a decrease of as a value by 0.01 Å2, after dimerization, it considered to be buried at PPI interface due to dimerization.

Surface Area of Interface

For a heterodimer the interface generally spans around > 600Å2 and for homodimer the area is larger. An interface surface area larger than 1000 Å2 , suggests an expected conformational changes of protomers after oligomerization.

Shape of the Interface

Studies from both experimental and computational field have shown approx 83% of interface is flat. That means any protruded residue should be complemented by an intruded residue. Also symmetry of the complex has been assumed to be correlated with this planarity of interface though the reverse is not true. As a gross look, each protomer has a small number residue protruding out from the surface and can be used to anchor physically into the partner, making the PPI stronger. A PPI interface can have two separate zone viz. core, buried in the interface, and rim, solvent accessible. The amino acid composition is also different for them.

Amino Acid Composition

Comparisons of interface amino acid residues with the other exterior part (surface) and interior (core) part of proteins reveal that, interface do possess some unique amino acid distribution than both surface and core of a protein. An overall increase in hydrophobic residues is expected than surface but core. Polar amino acids are somewhat comparable except N has higher affinity for interface and C is somewhat rare at PPI interface. Aromatic residues are generally common for interface. Among charged amino acids R is very likely to be found at PPI interface than core and surface respectively.

Distribution of Atom Types and Hydrophobicity of Interface

An increase affinity of non-polar atoms is also evident from studies compare to the other surface part. Thus a hydrophobic patch is more likely to be present at the interface, like a 'sticky plaster'. The distribution of these non-polar atoms is arranged in a concentric manner, resulting more or less a round zone of hydrophobic interaction.

Secondary Structure of Interface Residues

Comparatively the secondary structures (helix, beta sheet and loops like: turn and coil) of PPI interface are with a close similarity of protein surface, with beta sheet conformation being most common among them. Inside β arrangements more three types are found like extension (if at PPI interface protomers do form an arrangement like a continuous beta sheet between them), stacked (when protomers interact like they are lying on top of each other) and complex (neither type). Loop structures are also common after the beta sheet arrangement at the PPI interface.

Interface Discontinuity

PPI interface are not a continuous stretch of amino acid residues. They are present as surface patch of interaction. Mostly, residues present far than 5 residues are considered to be present in different patch.

Gap Volume and Surface Complementarity

A gap volume does indicate how close two protomer of a multimeric protein are allowed to come. Dividing the Gap volume (Å3) by interface ASA (Å2), we get gap volume index which shows most dimeric complexes, favoured by evolutionary selection, have a low value of Gap volume index, representing a strong interaction between them. Antigen antibody gap volume index have characteristic high value, showing a comparative weak interaction with foreign molecule. Surface complementarities of interface are measured by the gap volume (the gap or distance between two protomers).

Flexibility of Residue Amino Acids and Atoms at Interface

Flexibility of atoms or residues is considered to know if the Interface region is preformed or induced during oligomerization. Studies with relative ASA has shown the interface residues and atoms are more flexible than it have to be if they are preformed. Comparing crystallographic temperature factor (B factor), used to evaluate the displacement of an atom from its mean position of surface, core, and interface residue also have shown interface residues are more flexible than core but surface.

Forces Involved in PPI

For successful PPI, a fast but specific association is necessary. The functional oligomerization formation depends upon exact recognition of partners. Proper recognition has two essential qualities viz. affinity, cementing the PPI, and specificity, important for functional outcome. Some proteins do avail these criteria by co-localization and others have help from effector molecules. Several bonding, mainly non covalent, between protomers help them to hold each other.

Hydrogen Bonding

H bonding is a polar bond which arises due to polar interaction of two polar partners viz. a donor (having more electron affinity) and an acceptor (more electronegative). For oligomerization, side chain H bonds (both with main chain and side chain of partner) are more important than interaction between main chains of protomers.

Salt Bridges

Though not uncommon, salt bridges are relatively less in frequency at PPI interface. They are formed between oppositely charge amino acid residue interaction and if the residues are not far than 4 Å.

Disulphide Bonds

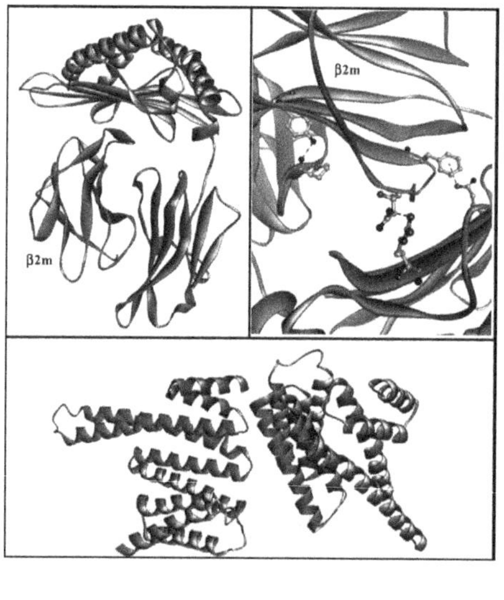

Figure shows (A) Major Histocompatibility Complex (MHC) I molecule (PDB ID: 3BO8). A heterodimer of A chain (A1 domain, ligand-bingding domain and A2 domain) and beta 2 microglobulin (β2m). (B) 14-3-3 beta isomer with homodimer of A and B chain (PDB ID: 2BQ0). C) Different bonds of PPI between β2m and A chains of MHC I molecule are shown. 1, A conventional H bond between proline and tyrosine; 2, salt bridge between arginine and aspartate; 3, π-cation interaction be-tween glutamine and phenyl alanine.

C has a low affinity for interface region, Disulphide bonds are very rare at PPI interface, particularly for intracellular protein and homo-complexes. Except these bonds, hydrophobic interaction plays a great role and is dominant interaction for oligomerization, particularly for permanent complexes.

Also, taxonomic comparison has revealed PPI interface are quite polar to exploit electrostatic interactions. Thus apart from salt bridges, also aromatic amino acids use their π electron cloud for interaction with cations of partners' side chain.

Methods to Study PPI

As it is now clear that more than 80% of proteins more commonly function in association rather in isolation, studying and understanding PPI is becoming more and more relevant for fields like system biology, molecular medicine etc. Studying PPI is also helpful to understand the molecular mechanisms behind mutations, related to disease onset and propagation. Several high throughput experimental data are present there and a comprehensive PPI network can be built. From studies of PPI we have many interesting information like:

1. Multifaceted nature of protein-protein complexes: Degeneracy is common for PPI interface structure, due to packing constraints, functional constraints or secondary structural elements of interface. PPI interface can be altered: Chemical alteration of PPI interface is possible and can be either beneficial or deleterious. This nature of PPI interface can be exploited for drug targeting or molecular medicine field. Mutations that cause a deleterious PPI interface alteration more likely to lead to a diseased condition.

2. Protein-protein complexes can be ordered and disordered: Transient protein complexes with weaker PPI interface are mostly stay in constant association-dissociation equilibrium. Thus disorderedness can be expected depending on its structural and functional requirements, in spite of entropic cost. Intrinsically disordered proteins (IDPs) and allosteric proteins or alternatively spliced signaling proteins do show same type of disordered interaction unlike permanent complexes that generally are more ordered after successful PPI.

3. Dynamic nature of Protein-protein complexes: Protein-protein complexes show flexibility like in enzyme catalytic domain or binding pockets for small molecules. Evolutionary changes do favor this dynamic nature. Protein engineering to sculpt PPI interface: this has shown a new way create a new interface for PPI, using a computational grafting of functional motifs onto a protein scaffold, thus making the interaction with more specificity and affinity. We might be benefitted by generating a protein with our desired functions.

For interaction fate, how a partner correctly finds the other among the large protein population of a cell, the whole protein surface is important as well as the PPI interface. Several ways are there to study PPI and the interacting interface. More commonly we can group them as in vivo, in vitro and in silico methods. The former two methods together form the experimental methods like in first type we express and colocalize protein inside a living body and the try to check the functions, like: Yeast two hybrid system, mutational analyses etc. In second type we do use a more controlled environment where we can modulate the protein molecules at per requirements like X-ray crystallography or NMR etc. Though these advance experimental studies are highly effective only a small fraction of expected number of protein-protein complexes in a living cell, has been studied due to time constraints, high costs and the evolution playing its best to incorporate new modifications into the interfaces of PPI. Thus, a third approach has started to get more importance now a day's viz. the in silico methods or computational methods where we can study, predict or exploit PPI interface at per our requirements, using the computers and simulation. In this third approach a good training dataset which has been made from the earlier experimental data, is mandatory and crucial as well as improvement of algorithms, playing important role in analyses.

Experimental Methods

Experimental studies of PPI involve studying the component subunit or protomers and mostly are done by comparing the molecular weights of both native protein and the subunits. There are two main strategies being followed for experimental studies.

In Vivo Studies

One of the most used techniques is the yeast two hybrid systems where we screen our protein of interest against a large random library of binding partner, using a marker gene expression after interaction. Yeast two hybrid system gives us an opportunity to study a direct interaction between proteins. But the main fallacies are that this method is prone to produce a large number of false positive and negative results. Interactions of proteins, not likely to be localized inside the nucleus, are difficult to be studied using this method. Proteins, requiring post-translational modifications and proper physiological conditions for correct folding and functioning, are also not suitable to study using this method. Another method here called Synthetic lethality, based on functional properties of associated complex, is also common practice for PPI study. As this method does not rely upon the physical interactions between proteins, allows only studying the functional interactions between proteins. Several other methods viz. bioluminescence energy transfers (BRET), fluorescence resonance energy transfer (FRET) and bimolecular fluorescence complementation (BiFC) are also used though requires extensive instrumentation.

In Vitro Studies

Studying PPI outside an organism is mostly limited to study the structural aspects by imaging or several other parameters. A main goal of imaging is to resolve the structure further better. Most of biophysical studies are limited to an ultimate objective of ab initio method for PPI interface prediction. There are a lot of them, though X-ray crystallography technique has been proved to produce most accurate results about PPI. This technique visualizes protein structures and the interacting complexes at atomic level, imaging like a high resolution microscopy. For this study precipitating a good crystal of protein from solution is of prime priority, making this technique less useful. As, for some integral membranc proteins like channels and receptors, an appropriate system to crystallize them is not very easy to create. Next to this comes NMR (Nuclear Magnetic Resonance Spectroscopy), which can detect protein structure in solution and also useful to detect weak PPIs. The fundamental knowledge of NMR lies beneath the fact that depending on the surrounding chemical environment, a magnetically active nucleus, induced by a strong magnetic field around it, can absorb electromagnetic wave of characteristic wavelength. The disadvantages of employing NMR arise due to the insensitivity of the method for which though atoms, present more frequently in the molecule, are more prominent and less frequent atoms might have dubious position due to faint signal and resolution. Also, working under high magnetic field affects many other instruments required during other studies make this method less user friendly. Motion sensitivity causing distortion of the image is also there. Affinity chromatography is another highly responsive method and able to detect weakest PPIs. But this method becomes less reliable due to its high specificity, causing a tendency of producing false positive result, detecting two proteins as interacting partners which might not be involved in PPI in cellular system. Tandem Affinity Purification-Mass Spectroscopy (TAP-MS) is another among in vitro techniques, based on marking proteins of interest at chromosomal locus and mass spectroscopy after double purification. Among other spectroscopic methods Matrix Assisted Laser Desorption Ionization-Time of Flight Mass Spectroscopy (MALDI-TOF MS) is also used to identify peptide ligands, using an affinity selection prior to spectroscopy. Optical Spectroscopy is also in line, using Fluorescence, highly sensitive to environmental changes. Circular Dichroism (CD), though having very limited application, is helpful to study PPI with secondary conformational changes. To identify, proteins involved in PPI from a whole cell extract Coimmuno precipitation [38] is very helpful. Using eukaryotic cells as an expression system has made this method enable to study the PPI of proteins which required extensive post translational modifications. Protein chips are used in protein microarrays technique that enable to test thousands of parameter within a single experiment. Protein fragment complementation assays (PCAs) can be used to detect PPI between hetero-oligomers.

Phage display is very helpful technique, using a phage virus incorporated with the protein and a gene of our interest. Isothermal titration calorimetry is based upon heat changes of solution due to protomer interactions. This method is helpful as does not require tagging or immobilizing the peptides. Among other techniques to study PPI, Surface Plasmon resonance (SPR) is also used for many purposes.

Computational Methods – Prediction of PPI

Though experimental methods are quite helpful to retrieve information about PPI interface, they has become not user friendly due to time consumption, high cost, requirements of high skill to perform and also not always reliable as all proteins do not form oligomer. Thus a prediction tool by computational method prior to any experimental trial is very useful to save time, money and human resources. Most of these methods use a training dataset with structural and sequence information of proteins, collected from earlier performed experimental studies. Some of well utilized parameters (variable) for PPI prediction studies are solvent accessibility, hydrophobicity, empirical energy functions, conservation score for amino acid residue etc. to name some of them. We can classify all types of PPI computational tools in broadly three categories i.e:

- Numerical value based methods: These methods rely on a function derived from the amino acid residue of our interest in protein chain, the information of neighboring residues and coefficients, determined by training. Now if the function value is above a certain cut off, we consider the residue under consideration either at interface or if not then at non interface. Sometimes, this approach is referred as Knowledge based approach. There are several methods in this approach viz. linear regression that establishes relationship among variables, scoring functions from known protein structure available from PDB. There are some machine learning tools as well to feed or train the computer with the training data set. Some of them are Support Vector Machines (SVM), Neural network and Random Forest, advantageous than others.

- Probabilistic methods: These methods mostly work based on conditional probability of whether a particular residue is at interface or not with the probability getting a particular value of a parameter of that residue. Among machine learning tools Naïve Bayesian has been well used. Bayesian network, Hidden Markov Model and Conditional Random Field are also shown with effective usage.
- Other biologically meaningful groups: PPI predictive methodologies have shown to use some biological points like evaluation based on whole genome analyses or evolutionary relationship or Proteins three dimensional structure, sequence, domains etc.

Each of above mention methods are in well use for both prokaryotes and eukaryotes. Predicting proteins functions are main aim of PPI, this revealing the heterogonous networks of proteins, resulting a better understanding of Proteomics and Interactomes.

Phenomics, Proteomics, PPI and Human Genetic Diseases

One among many driving forces for biological studies is to understand and find out solutions for human diseases, very important phenotypic character like health and evolutionary fitness. Though, studying only genetic variants and mutations did not suffice to understand how these complex biological mechanisms work. Today it has been found out that only a small fragment (only 3-5%) of our whole gene is worth studying. A huge surge for a need of more studies in phenomics is looming large, to understand the relation between diseases and genotypes of people, though creating phenome, a full set of phenotype of an individual, and then constructing a complete genotype-phenotype mapping is not an easy task. Phenotypic variations are not solely rely on genotypic variation and can have effective input from environment. A continuous interaction between genetic and environmental perturbations is thus influencing many of our polymorphic phenotypes as well as diseases. This complex relation (meaning of relation can be taken as a mapping of elements of one set to another set) between genotype and phenotype is making the task difficult as phenotypes are no longer functions (here, function stands for such a relation between genotype and phenotype set that each unique point in genotype set have exactly an unique point at phenotype set as outcome) of genotypes.

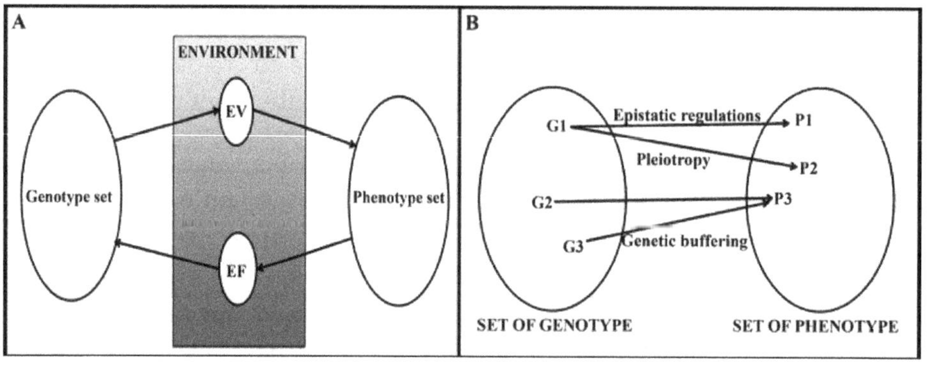

Figure shows (A) showing the variation in genotype of parents with environmental variations (EV) give rise to phenotypic variation of adult organisms. Evolutionary forces (EF) select best fit phenotype among them and thus shape the parental genotype for future generation. And the cycle moves on. (B) Showing how multiple genotypes can give rise to single phenotype (through genetic buffering) and a single genotype can regulate multiple phenotypes. Thus, a clear point to point mapping be-tween these two sets are very difficult.

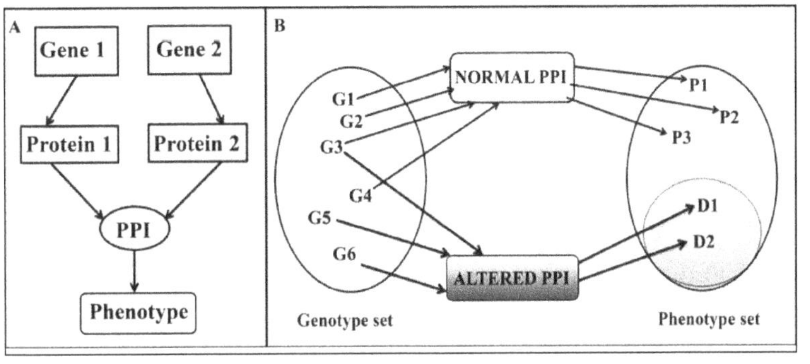

Figure phenotype (A) Showing, how a polygenic phenotype having input from more than one (here, the number is two) gene product is relying on the PPI of the proteins, coded from the genes, placing PPI at genotype-phenotype interface. (B) Showing how normal PPI and altered PPI (shown by thicker line) might be helpful for mapping genotypic variation with polymorphic phenotypic variations (P1, P2 ...) and diseases (D1, D2 ..., is presented here as subset of phenotype set).

For many diseases, a mutation in a single gene can have effects on multiple disorders or phenotypic outcome (genetic pleiotropy) and mutations in any single gene of a phenotype with multiple genetic interaction (locus heterogeneity) are contributing to this complexity of genotype-phenotype mapping (here, mapping is being considered as to find out an unique and exact point at phenotype set for each unique point at genotype set). Thus a genotype-phenotype mapping for genetic mutations and disease is in lack of success, having effects on therapeutic targeting.

Proteins do stand as prime mediator of our cellular phenotype by taking part in many, if not all, biochemical reactions or cell physiology and PPI comes to play a major role to complement the genetic data for diseases, affecting biological systems. It has been shown mutations inducing or correlated with genetic diseases are mostly located at the PPI interfaces. So we can visualize this like any genetic mutation can be treated as polymorphism if they are not altering the PPI and thus allowing the outcome or phenotype not to be evolutionary unfit than the wild type. While a mutation at PPI interfaces that can alter structure or can lead to an altered PPI (either due to loss of electrostatic interactions or hydrophobicity or steric hindrance etc.) has potential to alter the functions of the protein, the mutated gene product, and causing disease. Thus, assuming PPI at the interface of genotype-phenotype mapping would not be absolutely wrong and more rigorous research work on this field might be helpful for Phenomics, proteomics and disease related researches including therapeutic applications. Studies related to PPI interfaces' evolution has shown interfaces at permanent complex evolve more slowly, thus evolutionary more conserve, than interfaces involved in transient complex or weaker interactions. Disease causing mutations can be expected more to occur in a conserved domains. Disease causing mutations, altering PPI, can give rise to loss of interaction strength or position or unnecessary aggregation or unwanted changes during post translational modifications. Thus predicting the altered PPI due to any mutation at a particular locus might be helpful not only to predict the onset of a genetic disease but also could be exploited for drug designing research for genetic diseases. During host pathogen interaction it has been shown that pathogens and parasites do use a familiar PPI with host immune related protein to mimic host self-proteins, thus fleeing from host defense. For molecular medicine and medical biochemistry one of the most important research areas is drug targeting. Altering PPI has shown a huge potential for new medicine research. Targeting drugs to PPI interface though is not easy as it is a relatively large area. Thus more concentrated work on that field might be hopeful for new era of medicines.

Protein-Protein Interaction Detection

Protein-protein interaction detection methods are categorically classified into three types, namely, in vitro, in vivo, and in silico methods. In in vitro techniques, a given procedure is performed in a controlled environment outside a living organism. The in vitro methods in PPI detection are tandem affinity purification, affinity chromatography, coimmunoprecipitation, protein arrays, protein fragment complementation, phage display, X-ray crystallography, and NMR spectroscopy. In in vivo techniques, a given procedure is performed on the whole living organism itself. The in vivo methods in PPI detection are yeast two-hybrid (Y2H, Y3H) and synthetic lethality. In silico techniques are performed on a computer (or) via computer simulation. The in silico methods in PPI detection are sequence-based approaches, structure-based approaches, chromosome proximity, gene fusion, in silico 2 hybrid, mirror tree, phylogenetic tree, and gene expression-based approaches.

Table: Summary of PPI detection methods.

Approach	Technique	
In vitro	Tandem affinity purification-mass spectroscopy (TAP-MS) Affinity chromatography Coimmunoprecipitation Protein microarrays (H) Protein-fragment complementation Phage display (H) X-ray crystallography NMR spectroscopy	TAP-MS is based on the double tagging of the protein of interest on its chromosomal locus, followed by a two-step purification process and mass spectroscopic analysis. Affinity chromatography is highly responsive, can even detect weakest interactions in proteins, and also tests all the sample proteins equally for interaction. Coimmunoprecipitation confirms interactions using a whole cell extract where proteins are present in their native form in a complex mixture of cellular components. Microarray-based analysis allows the simultaneous analysis of thousands of parameters within a single experiment. Protein-fragment complementation assays (PCAs) can be used to detect PPI between proteins of any molecular weight and expressed at their endogenous levels. Phage-display approach originated in the incorporation of the protein and genetic components into a single phage particle. X-ray crystallography enables visualization of protein structures at the atomic level and enhances the understanding of protein interaction and function. NMR spectroscopy can even detect weak protein-protein interactions.

In vivo	Yeast 2 hybrid (Y2H) (H) Synthetic lethality	Yeast two-hybrid is typically carried out by screening a protein of interest against a random library of potential protein partners. Synthetic lethality is based on functional interactions rather than physical interaction.
In silico	Ortholog-based sequence approach Domain-pairs-based sequence approach Structure-based approaches Gene neighborhood Gene fusion In silico 2 hybrid (I2H) Phylogenetic tree Phylogenetic profile Gene expression	Ortholog-based sequence approach based on the homologous nature of the query protein in the annotated protein databases using pairwise local sequence algorithm. Domain-pairs-based approach predicts protein interactions based on domain-domain interactions. Structure-based approaches predict protein-protein interaction if two proteins have a similar structure (primary, secondary, or tertiary). If the gene neighborhood is conserved across multiple genomes, then there is a potential possibility of the functional linkage among the proteins encoded by the related genes. Gene fusion, which is often called as Rosetta stone method, is based on the concept that some of the single-domain containing proteins in one organism can fuse to form a multidomain protein in other organisms. The I2H method is based on the assumption that interacting proteins should undergo coevolution in order to keep the protein function reliable. The phylogenetic tree method predicts the protein-protein interaction based on the evolution history of the protein. The phylogenetic profile predicts the interaction between two proteins if they share the same phylogenetic profile. The gene expression predicts interaction based on the idea that proteins from the genes belonging to the common expression-profiling clusters are more likely to interact with each other than proteins from the genes belonging to different clusters.

In Vitro Techniques to Predict Protein-Protein Interactions

TAP tagging was developed to study PPIs under the intrinsic conditions of the cell. Gavin et al. first attempted the TAP-tagging method in a high-throughput manner to analyse the yeast interactome. This method is based on the double tagging of the protein of interest on its chromosomal locus, followed by a two-step purification process. Proteins that remain associated with the target protein can then be examined and identified through SDS-PAGE followed by mass spectrometry analysis, thereby identifying the PPI collaborator of the original protein of interest. An important dominance of TAP-tagging is its ability to identify a wide variety of protein complexes and to test the activeness of monomeric or multimeric protein complexes that exist in vivo. The TAP when used with mass spectroscopy (MS) will identify protein interactions and protein complexes.

The advantage of the affinity chromatography is that it is highly responsive, can even detect weakest interactions in proteins, and also tests all the sample proteins equally for interaction with the coupled protein in the column. However, false positive results also arise in the column due to high specificity among proteins, even though they do not get involved in the cellular system. Thus protein interaction studies cannot fully rely on affinity chromatography and hence require other methods in order to crosscheck and verify results obtained. The affinity chromatography can also be associated with SDS-PAGE technique and mass spectroscopy in order to generate a high-throughput data.

Coimmunoprecipitation confirms interactions using a whole cell extract where proteins are present in their native form in a complex mixture of cellular components that may be required for successful interactions. In addition, use of eukaryotic cells enables posttranslational modification which may be essential for interaction and which would not occur in prokaryotic expression systems.

Protein microarrays are rapidly becoming established as a powerful means to detect proteins, monitor their expression levels, and probe protein interactions and functions. A protein microarray is a piece of glass on which various molecules of protein have been affixed at separate locations in an ordered manner. Protein microarrays have seen tremendous progress and interest at the moment and have become one of the active areas emerging in biotechnology. The objective behind protein microarray development is to achieve efficient and sensitive high-throughput protein analysis, carrying out large numbers of determinations in parallel by automated process.

Protein-fragment complementation assay is another method of proteomics for the identification of protein-protein interactions in biological systems. Protein-fragment complementation assays (PCAs) are a family of assays for detecting protein-protein interactions (PPIs) that have been introduced to provide simple and direct ways to study PPIs in any living cell, multicellular organism, or in vitro. PCAs can be used to detect PPI between proteins of any molecular weight and expressed at their endogenous levels. The two choices for protein identification using a mass spectroscopy are peptide fingerprinting and shotgun proteomics. For peptide fingerprinting, the eluted complex is separated using SDS-PAGE. The gel is either Coomassie-stained or silver-stained and bands unique to the test sample and hopefully containing a single protein are excised, enzymatically digested, and analyzed by mass spectrometry. The mass of these peptides is determined and matched to a peptide database to determine the source protein. The gel also provides a rough estimate of the molecular weight of the protein. Since only unique bands are cut out, background bands are not identified. Abundant background proteins may obscure target proteins while less abundant proteins may fall below the limits of detection by staining. This method works well with purified samples containing only a handful of proteins. Alternatively, for shotgun proteomics, the entire eluate, containing many proteins, is digested. Shotgun proteomics is currently the most powerful strategy for analyzing such complicated mixtures.

There are different implementations of the phage display methodology as well as different applications. One of the most common approaches used is the M13 filamentous phage. The DNA encoding the protein of interest is ligated into the gene encoding one of the coat proteins of the virion. Normally, the process is followed by computational identification of potential interacting partners and a yeast two-hybrid validation step, but the method is a new born one.

X-ray crystallography is essentially a form of very high resolution microscopy, which enables visualization of protein structures at the atomic level and enhances the understanding of protein function. Specifically it shows how proteins interact with other molecules and the conformational changes in case of enzymes. Armed with this information, we can also design novel drugs that target a particular target protein.

In the recent past, the researchers have shown interest in the analysis of protein-protein interaction by nuclear magnetic resonance (NMR) spectroscopy. The location of binding interface is a crucial aspect in the protein interaction analysis. The basis for the NMR spectroscopy is that magnetically active nuclei oriented by a strong magnetic field absorb electromagnetic radiation at characteristic frequencies governed by their chemical environment.

In Vivo Techniques to Predict Protein-Protein Interactions

Y2H method is an *in vivo* method applied to the detection of PPIs. Two protein domains are required in the Y2H assay which will have two specific functions: (i) a DNA binding domain (DBD) that helps binding to DNA, and (ii) an activation domain (AD) responsible for activating transcription of DNA. Both domains are required for the transcription of a reporter gene. Y2H analysis allows the direct recognition of PPI between protein pairs. However, the method may incur a large number of false positive interactions. On the other hand, many true interactions may not be traced using Y2H assay, leading to false negative results. In an Y2H assay, the interacting proteins must be localized to the nucleus, since proteins, which are less likely to be present in the nucleus are excluded because of their inability to activate reporter genes. Proteins, which need posttranslational modifications to carry out their functions, are unlikely to behave or interact normally in an Y2H experiment. Furthermore, if the proteins are not in their natural physiological environment, they may not fold properly to interact. During the last decade, Y2H has been enriched by designing new yeast strains containing multiple reporter genes and new expression vectors to facilitate the transformation of yeast cells with hybrid proteins. Other widely used techniques, such as bioluminescence resonance energy transfer (BRET), fluorescence resonance energy transfers (FRET), and bimolecular fluorescence complementation (BiFC), require extensive instrumentation. FRET uses time-correlated single-photon counting to predict protein interactions.

Synthetic lethality is an important type of *in vivo* genetic screening which tries to understand the mechanisms that allow phenotypic stability despite genetic variation, environmental changes, and random events such as mutations. This methodology produces mutations or deletions in two or more genes which are viable alone but cause lethality when combined together under certain conditions. Compared with the results obtained in the aforesaid methods, the relationships detected by synthetic lethality do not require necessity of physical interaction between the proteins. Therefore, we refer to this type of relationships as functional interactions.

In Silico Methods for the Prediction of Protein-Protein Interactions

The yeast two-hybrid (Y2H) system and other in vitro and in vivo approaches resulted in large-scale development of useful tools for the detection of protein-protein interactions (PPIs) between specified proteins that may occur in different combinations. However, the data generated through these approaches may not be reliable because of non-availability of possible PPIs. In order to understand the total context of potential interactions, it is better to develop approaches that predict the full range of possible interactions between proteins.

A variety of in silico methods have been developed to support the interactions that have been detected by experimental approach. The computational methods for in silico prediction include sequence-based approaches, structure-based approaches, chromosome proximity, gene fusion, in silico 2 hybrid, mirror tree, phylogenetic tree, gene ontology, and gene expression-based approaches. The list of all webservers of in silico methods was given in Table below.

Table: The list of web servers.

S. No.	Web server	Function
1	Struct2Net	The Struct2Net server makes structure-based computational predictions of protein-protein interactions (PPIs).
2	Coev2Net	Coev2Net is a general framework to predict, assess, and boost confidence in individual interactions inferred from a high-throughput experiment.
3	PRISM PROTOCOL	PRISM PROTOCOL is a collection of programs that can be used to predict protein-protein interactions using protein interfaces.
4	InterPreTS	InterPreTS uses tertiary structure to predict interactions.
5	PrePPI	PrePPI predicts protein interactions using both structural and nonstructural information.
6	iWARP	iWARP is a threading-based method to predict protein interaction from protein sequences.
7	PoiNet	PoiNet provides PPI filtering and network topology from different databases.
8	PreSPI	PreSPI predicts protein interactions using a combination of domains.
9	PIPE2	PIPE2 queries the protein interactions between two proteins based on specificity and sensitivity.
10	HomoMINT	HomoMINT predicts interaction in human based on ortholog information in model organisms.
11	SPPS	SPPS searches protein partners of a source protein in other species.
12	OrthoMCL-DB	OrthoMCL-DB is a graph-clustering algorithm designed to identify homologous proteins based on sequence similarity.
13	P-POD	P-POD provides an easy way to find and visualize the orthologs to a query sequence in the eukaryotes.
14	COG	COG shows phylogenetic classification of proteins encoded in genomes.
15	BLASTO	BLASTO performs BLAST based on ortholog group data.
16	PHOG	PHOG web server identifies orthologs based on precomputed phylogenetic trees.
17	G-NEST	G-NEST is a gene neighborhood scoring tool to identify co-conserved, coexpressed genes.
18	InPrePPI	InPrePPI predicts protein interactions in prokaryotes based on genomic context.
19	STRING	Containing both physical and functional associations.
20	MirrorTree	The MirrorTree allows graphical and interactive study of the coevolution of two protein families and asseses their interactions in a taxonomic context.
21	TSEMA	TSEMA predicts the interaction between two families of proteins based on Monte Carlo approach.

Structure-based Prediction Approaches

The idea behind the structure-based method is to predict protein-protein interaction if two proteins have a similar structure. Therefore, if two proteins A and B can interact with each other, then there may be two other proteins and whose structures are similar to those of proteins A and B; then it is implied that proteins and can also interact with each other. But most proteins may not be having known structures; the first step for this method is to guess the structure of the protein based on its sequence. This can be done in different ways. The PDB database offers useful tools and information resources for researchers to build the structure for a query protein. Using the multimeric threading approach, Lu et al. have made 2,865 protein-protein interactions in yeast and 1,138 interactions have been confirmed in the DIP.

Recently, Hosur et al. developed a new algorithm to infer protein-protein interactions using structure-based approach. The Coev2Net algorithm, which is a three-step process, involves prediction of the binding interface, evaluation of the compatibility of the interface with an interface coevolution based model, and evaluation of the confidence scores for the interaction. The algorithm when applied to binary protein interactions has boosted the performance of the algorithm over existing methods. However, Zhang et al. have used three-dimensional structural information to predict PPIs with an accuracy and coverage that are superior to predictions based on nonstructural evidence.

Sequence-based Prediction Approaches

Predictions of PPIs have been carried out by integrating evidence of known interactions with information regarding sequential homology. This approach is based on the concept that an interaction found in one species can be used to infer the interaction in other species. However, recently, Hosur et al. developed a new algorithm to predict protein-protein interactions using threading-based approach which takes sequences as input. The algorithm, iWARP (Interface Weighted RAPtor), which predicts whether two proteins interact by combining a novel linear programming approach for interface alignment with a boosting classifier for interaction prediction. Guilherme Valente et al. introduced a new method called Universal *In Silico* Predictor of Protein-Protein Interactions (UNISPPI), based on primary sequence information for classifying protein pairs as interacting or non-interacting proteins . Kernel methods are hybrid methods which use a combination of properties like protein sequences, gene ontologies, and so forth. However, there are two different methods under sequence-based criterion:

- Ortholog-Based Approach*:* The approach for sequence-based prediction is to transfer annotation from a functionally defined protein sequence to the target sequence based on the similarity. Annotation by similarity is based on the homologous nature of the query protein in the annotated protein databases using pairwise local sequence algorithm. Several proteins from an organism under study may share significant similarities with proteins involved in complex formation in other organisms.

The prediction process starts with the comparison of a probe gene or protein with those annotated proteins in other species. If the probe gene or protein has high similarity to the sequence of a gene or protein with known function in another species, it is assumed that the probe gene or protein has either the same function or similar properties. Most subunits of protein complexes were annotated in that way. When the function is transferred from a characterized protein to an uncharacterized protein, ortholog and paralog concepts should be applied. Orthologs are the genes in different species that have evolved from a common ancestral gene by speciation. In contrast, paralogs usually refer to the genes related by duplication within a genome. In broad sense, orthologs will retain the functionality during the course of evolution, whereas paralogs may acquire new functions. Therefore, if two proteins—A and B—interact with each other, then the orthologs of A and B in a new species are also likely to interact with each other:

- Domain-Pairs-Based Approach: A domain is a distinct, compact, and stable protein structural unit that folds independently of other such units. But most of times, domains are defined as distinct regions of protein sequence that are highly conserved in the process of evolution. As individual structural and functional units, protein domains play an important role in the development of protein structural class prediction, protein subcellular location prediction, membrane protein type prediction, and enzyme class and subclass prediction.

Conventionally, protein domains are used for basic research and also for structure-based drug designing. In addition, domains are directly involved in the intermolecular interaction and hence must be fundamental to protein-protein interaction. Multiple studies have shown that domain-domain interactions (DDIs) from different experiments are more consistent than their corresponding PPIs. So, it is quite reliable to use the domains and their interactions for prediction of the protein-protein interactions and vice versa.

Chromosome Proximity/Gene Neighbourhood

With the ever increasing number of the completely sequenced genomes, the global context of genes and proteins in the completed genomes has provided the researchers with the enriched information needed for the protein-protein interaction detection. It is well known that the functionally related proteins tend to be organized very closely into regions on the genomes in prokaryotes, such as operons, the clusters of functionally related genes transcribed as a single mRNA. If the neighborhood relationship is conserved across multiple genomes, then it will be more relevant for implying the potential possibility of the functional linkage among the proteins encoded by the related genes. And this evidence was applied to study the functional association of the corresponding proteins. This relationship was confirmed by the experimental results and shown to be more independent of relative gene orientation. Recently, it has been found that there is functional link among the adjacent bidirectional genes along the chromosome. Interestingly, in most cases, the relationship among adjacent bidirectionally transcribed genes with conserved gene orientation is that one gene encodes a transcriptional regulator and the other belongs to non-regulatory protein. It has been found that most of the regulators control the transcription of the diver gently transcribed target gene/operon and automatically regulate their own biosynthesis as well. This relationship provides another way to predict the target processes and regulatory features for transcriptional regulators. One of the pitfalls of this method is that it is directly suitable for bacterial genome since gene neighboring is conserved in the bacteria.

Gene Fusion

Gene fusion, which is often called as Rosetta stone method, is based on the concept that some of the single-domain containing proteins in one organism can fuse to form a multidomain protein in other organisms. This domain fusion phenomenon indicates the functional association for those separate proteins, which are likely to form a protein complex. It has been shown that fusion events are particularly common in those proteins participating in the metabolic pathway. This method can be used to predict protein-protein interaction by using information of domain arrangements in different genomes. However, it can be applied only to those proteins in which the domain arrangement exists.

In Silico Two-Hybrid (I2h)

The method is based on the assumption that interacting proteins should undergo coevolution in order to keep the protein function reliable. In other words, if some of the key amino acids in one protein changed, the related amino acids in the other protein which interacts with the mutated counter partner should also make the compulsory mutations as well. During the analysis phase, the common genomes containing those two proteins will be identified through multiple sequence alignments and a correlation coefficient will be calculated for every pair of residues in the same protein and between the proteins. Accordingly, there are three different sets for the pairs: two from the intraprotein pairs and one from the interprotein pairs. The protein-protein interaction is inferred based on the difference from the distribution of correlation between the interacting partners and the individual proteins. Since I2h analysis is based on the prediction of physical closeness between residue pairs of the two individual proteins, the result from this method automatically indicates the possible physical interaction between the proteins.

Phylogenetic Tree

Another important method for detection of interaction between the proteins is phylogenetic tree. The phylogenetic tree gives the evolution history of the protein. The mirror tree method predicts protein-protein interactions under the belief that the interacting proteins show similarity in molecular phylogenetic tree because of the coevolution through the interaction. The underlying principle behind the method is that the coevolution between the interacting proteins can be reflected from the degree of similarity from the distance matrices of corresponding phylogenetic trees of the interacting proteins. The set of organisms common to the two proteins are selected from the multiple sequence alignments (MSA) and the results are used to construct the corresponding distance matrix for each protein. The BLAST scores could also be used to fill the matrices. Then the linear correlation is calculated among these distance matrices. High correlation scores indicate the similarity between the phylogenetic trees and therefore the proteins are considered to have the interaction relationship. The MirrorTree method is used to detect the coevolution relationship between proteins and the results are used to infer the possibility of their physical interaction.

Phylogenetic Profile

The notion for this method is that the functionally linked proteins tend to coexist during the evolution of an organism. In other words, if two proteins have a functional linkage in a genome, there will be a strong pressure on them to be inherited together during evolution process. Thus, their corresponding orthologs in other genome will be preserved or dropped. Therefore, we can detect the presence or absence (co-occurrence) of proteins in the phylogenetic profile. A phylogenetic profile describes an occurrence of a certain protein in a set of genomes: if two proteins share the same phylogenetic profiling, this indicates that the two proteins have the functional linkage. In order to construct the phylogenetic profile, a predetermined threshold of BLASTP -value is used to detect the presence or absence of the homological proteins on the target genome with the source genomes. This method gives promising results in the detection of the functional linkage among the proteins and, at the same time, assigns the functions to query proteins. Even though the phylogenetic profile has shown great potential for building the functional linkage network on the full genome level, the following two pitfalls should be mentioned: one is that this method is based on full genome sequences and the other is that the functional linkage between proteins is detected by their phylogenetic profiling, so it is difficult to use the method for those essential proteins in the cell where no difference can be detected from the phylogenetic profile. Moreover, even though the increasing number in the source genome set can improve the prediction accuracy, there may be an upper limit for this method.

Many genomic events contribute to the noise during the coevolution, such as gene duplication or the possible loss of gene functions in the course of evolution, which could corrupt the phylogenetic profile of single genes. Phylogenetic-profile-based methods conceded satisfactory performance only on prokaryotes but not on eukaryotes.

Gene Expression

The method takes the advantage of high-throughput detection of the whole gene transcription level in an organism. Gene expression means the quantification of the level at which a particular gene is expressed within a cell, tissue or organism under different experimental conditions and time intervals. By applying the clustering algorithms, different expression genes can be grouped together according to their expression levels, and the resultant gene expression under different experimental conditions can help to enunciate the functional relationships of the various genes. A lot of research has also been carried out to investigate the relationship between gene co-expression and protein interaction. Based on the yeast expression data and proteome data, proteins from the genes belonging to the common expression-profiling clusters are more likely to interact with each other than proteins from the genes belonging to different clusters. In other studies, it has been confirmed that adjacent genes tend to be expressed both in the eukaryotes and prokaryotes. The gene co-expression concept is an indirect way to infer the protein interaction, suggesting that it may not be appropriate for accurate detection of protein interactions. However, as a complementary approach, gene co-expression can be used to validate interactions generated from other experimental methods.

Comparison of Protein-Protein Interaction Methods

Each of the above methods has been applied to detect the protein-protein interaction in both the prokaryotes and eukaryotes. The results show that most of them fit better for the prokaryotes than eukaryotes. The significant increase for the coverage among those studies during the past several years could be mainly because of the increase in the number of genomes being decoded. This is because the more the number of genomes used in the study, the higher the coverage that the methods can reach. With the accumulation of fully sequenced genomes, the information content in the reference genome set is expected to increase. Accordingly, the prediction accuracy would increase with more genomes incorporated in the study. It can be anticipated that, with more and more genomes available in the future, the prediction potential will be improved and the corresponding combined methods will get higher coverage and accuracy. One thing that should be mentioned is that the selection of the standard used for the evaluation of the methods has a great impact on the coverage and accuracy. Besides the Operon and Swiss-Prot key word recovery used in the above studies, the KEGG has been used as the standard in Search Tool for the Retrieval of Interacting Genes (STRING) database. It can be expected that the prediction coverage and accuracy will be different for each method under different standards. Obviously, the achieved highest coverage for the gene order method based on the operon standard indicates that the method is strongly related to operon.

Recent technological advances have allowed high-throughput measurements of protein-protein interactions in the cell, producing protein interaction networks for different species at a rapid pace. However, high-throughput methods like yeast two-hybrid, MS, and phage display have experienced high rates of noise and false positives. There are some verification methods to know the reliability of these high throughput interactions. They are Expression Profile Reliability (EPR index), Paralogous Verification Method (PVM) Protein Localization Method (PLM) [60], and Interaction Generalities Measures IG1 and IG2. EPR method compares protein interaction with RNA expression profiles whereas PVM analyzes paralogs of interactors for comparison. The IG1 measure is based on the idea that interacting proteins that have no further interactions beyond level-1 are likely to be false positives. The IG2 measure uses the topology of interactions. Bayesian approaches have also been used for calculation of reliability. The PLM gives the true positives (TP) as interacting proteins, which need to be localized in the same cellular compartment or annotated to have a common cellular role. So, in order to counter these errors, many methods have been developed which provides confidence scores with each interaction. Also, the methods that assign scores to individual interactions generally perform better than those with the set of interactions obtained from an experiment or a database.

Computational Analysis of PPI Networks

A PPI network can be described as a heterogeneous network of proteins joined by interactions as edges. The computational analysis of PPI networks begins with the illustration of the PPI network arrangement. The simplest sketch takes the form of a mathematical graph consisting of nodes and edges. Protein is represented as a node in such a graph and the proteins that interact with it physically are represented as adjacent nodes connected by an edge. An examination of the network can yield a variety of results. For example, neighboring proteins in the graph probably may share more the same functionality. In addition to the functionality, densely connected sub-graphs in the network are likely to form protein complexes as a unit in certain biological processes. Thus, the functionality of a protein can be inferred by spotting at the proteins with which it interacts and the protein complexes to which it resides. The topological prediction of new interactions is a novel and useful option based exclusively on the structural information provided by the PPI network (PPIN) topology. Some algorithms like random layout algorithm, circular layout algorithm, hierarchical layout algorithm, and so forth are used to visualize the network for further analysis. Precisely, the computational analysis of PPI networks is challenging, with these major barriers being commonly confronted:

- The protein interactions are not stable.
- One protein may have different roles to perform.
- Two proteins with distinct functions periodically interact with each other.

Role of PPI Networks in Proteomics

Predicting the protein functionality is one of the main objectives of the PPI network. Despite the recent comprehensive studies on yeast, there are still a number of functionally unclassified proteins in the yeast database which reflects the impending need to classify the proteins. The functional annotation of human proteins can provide a strong foundation for the complete understanding of cell mechanisms, information that is valuable for drug discovery and development. The increased availability of PPI networks has developed various computational methods to predict protein functions. The availability of reliable information on protein interactions and their presence in physiological and pathophysiological processes are critical for the development of protein-protein-interaction-based therapeutics. The compendium of all known protein-protein interactions (PPIs) for a given cell or organism is called the interactome.

Protein functions may be predicted on the basis of modularization algorithms. However, predictions found in this way may not be accurate because the accuracy of the modularization process itself is typically low. There are other methods which include the neighbor counting, Chi-square, Markov random field, Prodistin, and weighted-interactions-based method for the prediction of protein function. For greater accuracy, protein functions should be predicted directly from the topology or connectivity of PPI networks. Several topology-based approaches that predict protein function on the basis of PPI networks have been introduced. At the simplest level, the "neighbor counting method" predicts the function of an unexplored protein by the frequency of known functions of the immediate neighbor proteins. The majority of functions of the immediate neighbors can be statistically assessed. Recently, the number of common neighbors of the known protein and the unknown protein has been taken as the basis for the inference of function. The weighted-graph-mining-based protein function prediction is a novel approach in the area.

Protein-protein complex identification is the crucial step in finding the signal transduction pathways. Protein-protein complexes mostly consist of antibody-antigen and protease-inhibitor complexes. Crystallography is the major tool for determining protein complexes at atomic resolution.

The complete analysis of PPIs can enable better understanding of cellular organization, processes, and functions. The other applications of PPI Network include biological indispensability analysis, assessing the drug ability of molecular targets from network topology, estimation of interactions reliability, identification of domain-domain interactions, prediction of protein interactions, detection of proteins involved in disease pathways, delineation of frequent interaction network motifs, comparison between model organisms and humans, and protein complex identification.

Protein Interaction Databases

The massive quantity of experimental PPI data being generated on steady basis has led to the construction of computer-readable biological databases in order to organize and to process this data. For example, the biomolecular interaction network database (BIND) is created on an extensible specification system that permits an elaborate description of the manner in which the PPI data was derived experimentally. The database of interacting proteins (DIP) is another database of experimentally determined protein-protein binary interactions. The biological general repository for interaction datasets (BioGRID) is a database that contains protein and genetic interactions among thirteen different species. Interactions are regularly added through exhaustive curation of the primary literature to the databases. Interaction data is extracted from other databases including BIND and MIPS (Munich Information Center for protein se quences), as well as directly from large-scale experiments. HitPredict is a resource of high confidence protein-protein interactions from which we can get the total number of interactions in a species for a protein and can view all the interactions with confidence scores.

The Molecular Interaction (MINT) database is another database of experimentally derived PPI data extracted from the literature, with the added element of providing the weight of evidence for each interaction. The Human Protein Interaction Database (HPID) was designed to provide human protein interaction information precomputed from existing structural and experimental data. The information Hyperlinked over Proteins (iHOP) database can be searched to identify previously reported interactions in PubMed for a protein of interest. IntAct provides an open source database and toolkit for the storage, presentation, and analysis of protein interactions. The web interface provides both textual and graphical representations of protein interactions and allows exploring interaction networks in the context of the GO annotations of the interacting proteins. However, we have observed that the intersection and overlap between these source PPI databases is relatively small. Recently, the integration has been done and can be explored in the web server called APID (Agile Protein Interaction Data Analyzer) which is an interactive bioinformatics' web tool developed to allow exploration and analysis of currently known information about protein-protein interactions integrated and unified in a common and comparative platform . The Protein Interaction Network Analysis (PINA2.0) platform is a comprehensive web resource, which includes a database of unified protein-protein interaction data integrated from six manually curated public databases and a set of built-in tools for network construction, filtering, analysis, and visualization. The databases and number of interactions were tabled in Table below.

Table: Protein interaction databases.

S. No.	Database name	Total number of interactions (to date)	Number of species/organisms
1	BioGrid	7,17,604	60
2	DIP	76,570	637
3	HitPredict	2,39,584	9
4	MINT	2,41,458	30
5	IntAct	4,33,135	8
6	APID	3,22,579	15
7	BIND	>3,00,000	–
8	PINA2.0	3,00,155	7

Covalent Bond

A covalent bond is formed by equal sharing of electrons from both the participating atoms. The pair of electrons participating in this type of bonding is called shared pair or bonding pair. The covalent bonds are also termed as molecular bonds. Sharing of bonding pairs will ensure that the atoms achieve stability in their outer shell which is similar to the atoms of noble gases.

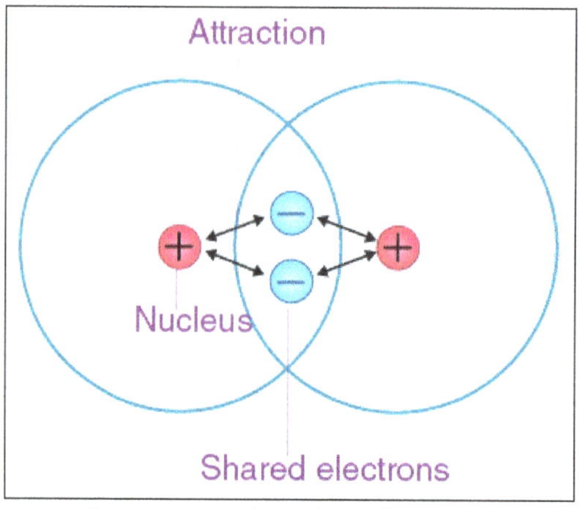

Figure: Formation of Covalent Bond.

Elements having very high ionisation energies are incapable of transferring electrons and elements having very low electron affinity cannot take up electrons. The atoms of such elements tend to share their electrons with the atoms of other elements or with other atoms of the same element in a way that both the atoms obtain octet configuration in their respective valence shell and thus achieve stability. Such association through sharing of electron pairs among different or same kinds is known as Covalent Bond. Covalent Bonding can be achieved in two Ways:

- Sharing of electrons between atoms of the same kind E.g. Formation of H_2, Cl_2, O_2 etc.
- Sharing of electrons between atoms of different kind E.g. Formation of CH_4, H_2O, NH_3 etc.

Covalent Bonding in Carbon Atom

As per the electronic configuration of Carbon, it needs to gain or lose 4 electrons to become stable, which seems impossible as:

- Carbon cannot gain 4 electrons to become C4-, because it will be tough for 6 protons to hold 10 electrons and so the atom will become unstable.
- Carbon cannot lose 4 electrons to become C4+ because it would require a large amount of energy to remove out 4 electrons and also the C4+ would have only 2 electrons held by proton, which will again become unstable.

Carbon cannot gain or donate electrons, so to complete its nearest noble gas configuration, it shares electron to form a covalent bond.

Properties of Covalent Bond

If the normal valence of an atom is not satisfied by sharing a single electron pair between atoms, the atoms may share more than one electron pair between them. Some of the properties of covalent bonds are:

- Covalent bonding does not result in the formation of new electrons. The bond only pairs them.

- They are very powerful chemical bonds that exist between atoms.
- A covalent bond normally contains the energy of about ~80 kilocalories per mole (kcal/mol).
- Covalent bonds rarely break spontaneously after it is formed.
- Covalent bonds are directional where the atoms that are bonded showcase specific orientations relative to one another.
- Most compounds having covalent bonds exhibit relatively low melting points and boiling points.
- Compounds with covalent bonds usually have lower enthalpies of vaporization and fusion.
- Compounds formed by covalent bonding don't conduct electricity due to the lack of free electrons.
- Covalent compounds are not soluble in water.

Octet Rule

All atoms except noble gases have less than eight electrons in their valence shell. In other words, the valence shells of these atoms do not have stable configurations. Therefore, they combine with each other or with other atoms to attain stable electronic configurations. Therefore,

"The tendency of atoms of various elements to attain stable configuration of eight electrons in their valence shells is the cause of Chemical combination" and "The principle of attaining the maximum of eight electrons in the valence shell of atoms is called octet rule."

Lewis introduced simple symbols to denote the electrons present in the outer shell of atom known as the valence electrons. These symbols are known as Electron Dot Symbols and the structure of the compound is known as Lewis Dot Structure.

Figure: Dot structure of methane.

Conditions for Writing the Lewis Dot Structures

- Sharing of an electron pair between the atoms results in the formation of covalent bonds.
- During bond formation, each bond consists of two electrons which are contributed by each one of the combining atoms.
- By the mutual sharing of electrons, each atom attains octet configuration in its valence shell.

Electron dot structures of covalent molecules are written with respect to the octet rule. According to this rule, all the atoms in the molecule will have eight electrons in their valence shell except the Hydrogen atom. Hydrogen will have only two electrons because only two electrons complete its first shell to attain helium configuration. Thus the elements of group 17 such as Cl would share one electron to attain stable octet; The elements of group 16 such as O and S would share two electrons; the elements of group 15 would share three electrons and so on.

For Example, the oxygen atom which has six electrons in its valence shell completes its octet by sharing its two electrons with two hydrogen atoms to form a water molecule.

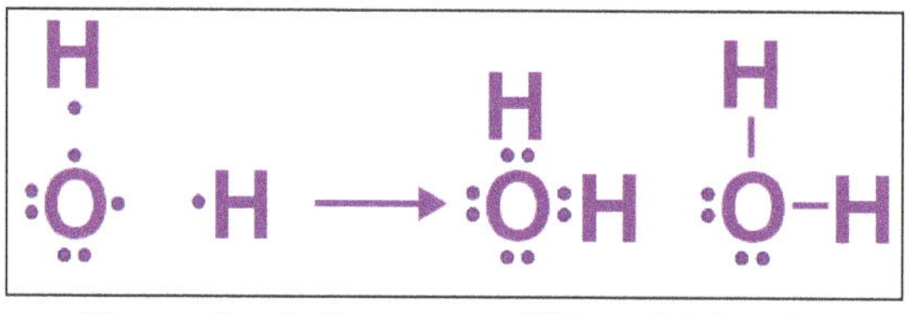

Figure: Lewis Structure of Water Molecule.

Types of Covalent Bonds

Depending upon the number of shared electron pairs, the covalent bond can be classified into:

- Single Covalent Bond.
- Double Covalent Bond.
- Triple Covalent Bond.

Single Bonds

A single bond is formed when only one pair of the electron is shared between the two participating atoms. It is represented by one dash (-). Although this form of covalent bond has a smaller density and is weaker than a double and triple bond, it is the most stable.

For Example, HCL molecule has one Hydrogen atom with one valence electron and one Chlorine atom with seven valence electrons. In this case, a single bond is formed between hydrogen and chlorine by sharing one electron.

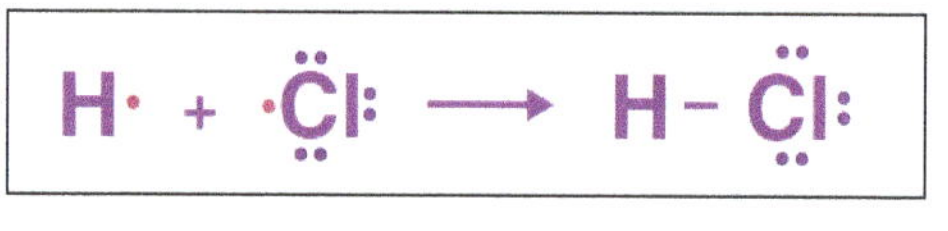

Double Bonds

A double bond is formed when two pairs of electrons are shared between the two participating atoms. It is represented by two dashes (=). Double covalent bonds are much stronger than a single bond, but they are less stable.

Example: Carbon dioxide molecule has one carbon atom with six valence electrons and two oxygen atom with four valence electrons.

To complete its octet, carbon shares two of its valence electrons with one oxygen atom and two with another oxygen atom. Each oxygen atom shares its two electrons with carbon and therefore there are two double bonds in CO_2.

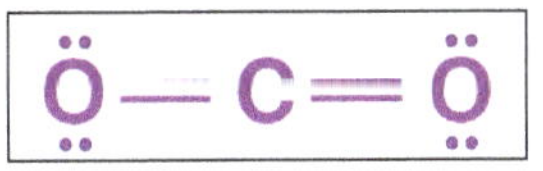

Figure: CO_2 Molecule with Double Covalent Bond.

Oxygen-Molecule: In the formation of the oxygen molecule, each oxygen atom has six electrons in their valence shell. Each atom requires two more electrons to complete their octet. Therefore the atoms share two electrons each to form the oxygen molecule. Since two electron pairs are shared there is a double bond between the two oxygen atoms.

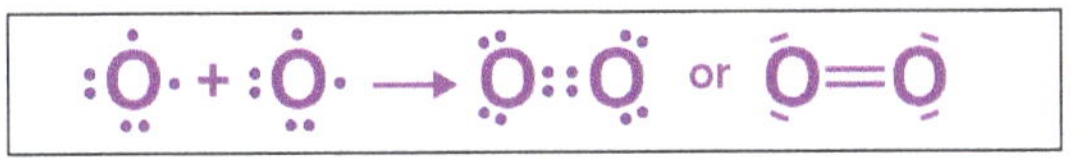

Figure: O_2 Molecule with Double Covalent Bond.

Ethylene Molecule: In ethylene, each carbon atom shares two of its valence electron with two hydrogen atoms and remaining two electrons with the other carbon atom. So there is a double bond between the carbon atoms.

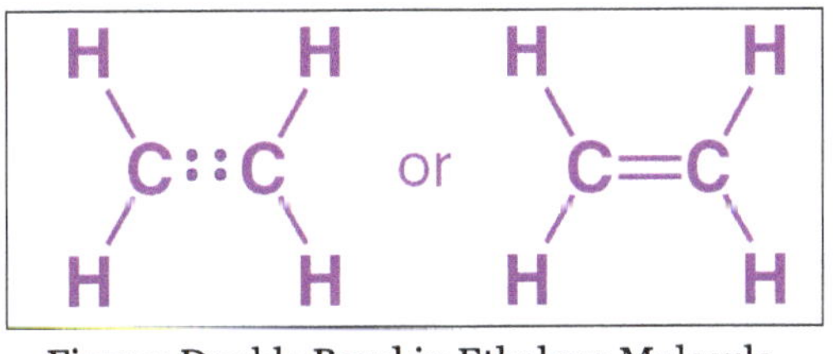

Figure: Double Bond in Ethylene Molecule.

Triple Bond

A triple bond is formed when three pairs of electrons are shared between the two participating atoms. Triple covalent bonds are represented by three dashes (≡) and are the least stable types of covalent bonds.

For Example: In the formation of a nitrogen molecule, each nitrogen atoms having five valence electrons provides three electrons to form three electron pairs for sharing. Thus, a triple bond is formed between the two nitrogen atoms.

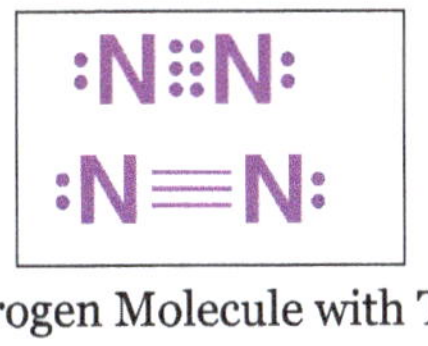

Figure: Nitrogen Molecule with Trple Bond.

Polar Covalent Bond

This type of covalent bond exists where the unequal sharing of electrons occurs due to the difference in the electronegativity of combining atoms. More electronegative atom will have a stronger pull for electrons. The electronegative difference between the atoms is greater than zero and less than 2.0. As a result, the shared pair of electrons will be closer to that atom.

Example: Molecules forming hydrogen bonding as a result of an unbalanced electrostatic potential. In this case, the hydrogen atom interacts with electronegative fluorine, hydrogen, or oxygen.

Nonpolar Covalent Bond

This type of covalent bond is formed whenever there is an equal share of electrons between atoms. The electronegativity difference between two atoms is zero. It occurs wherever the combining atoms have similar electron affinity (diatomic elements).

Example: Nonpolar Covalent Bond is found in gas molecules like Hydrogen gas, Nitrogen gas, etc.

Polarization of Covalent Bonds

It is observed that in the sigma bonds between two different atoms, the electron cloud is always closer to the more electronegative of the two atoms participating in the sigma bond. Due to this, there is a permanent dipole that arises in the bond and the covalent bond is said to be polarized.

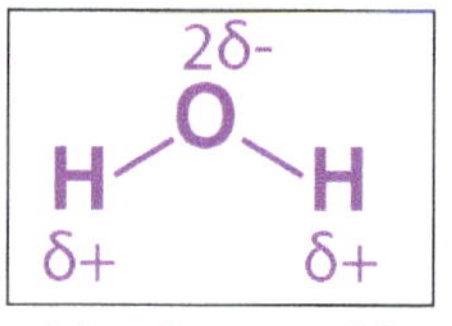

Figure: Polarity of Covalent Bond in water Molecule.

An illustration describing the polarity of the covalent bonds in a water molecule is provided above. The more electronegative atom is said to have a partial negative charge and the less electronegative atom has a partial positive charge in the polar covalent bond.

Difference between Covalent and Ionic Bonds

Covalent bonds and ionic bonds are types of atomic bonds. These bonds are different in their properties and structure. The covalent bonds include pairs of electrons by two atoms binding them in a fixed orientation. While a bond between two ions is called ionic bonds.

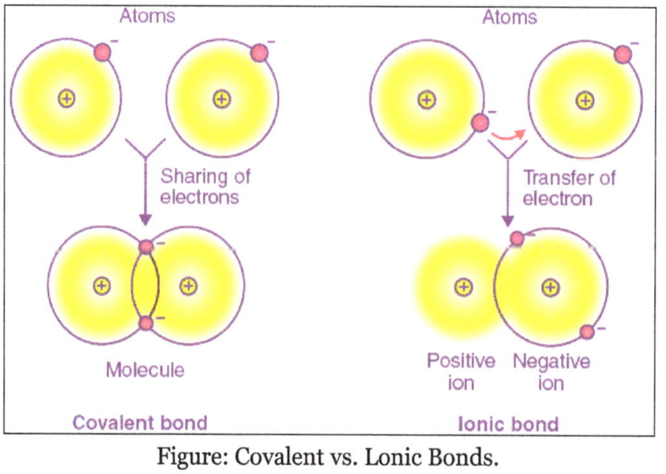

Figure: Covalent vs. Lonic Bonds.

Covalent bonding occurs between two non-metallic atoms characterized by the sharing of electron pairs between the atoms and other covalent bonds with electronegativity difference is greater than 2.0 (<2.0). In the case of covalent bond formation, polyatomic ions are formed. Whereas, the ionic bond is formed as a result of electrostatic attraction between the oppositely charged ions.

Non-Covalent Bond

A noncovalent bond is a type of chemical bond that typically bond between macromolecules. They do not involve sharing a pair of electrons. Noncovalent bonds are used to bond large molecules such as proteins and nucleic acids. Noncovalent bonds are weaker than covalent bonds but they are crucial for biochemical processes such as the formation of double helix. There are four commonly mentioned fundamental noncovalent bond types. They include electrostatic interactions, hydrogen bonds, van der Waals interactions, and hydrophobic interactions. Each type differs in geometry, strength, and specificity.

Types of Noncovalent Bonds

Electrostatic Interactions

The energy of an electrostatic interaction is given by Coulomb's Law: $E = kq_1q_2 / Dr^2$ where E is the energy, q_1 and q_2 are the charges of two atoms, r is the distance between the two atoms, D is the dielectric constant, and k is a proportionality constant. k = 1389 for energy that's measured in kilojoules/mol or k = 332 for energy that's measured in kcal/mol. A charged group on a molecule can attract an oppositely charged group from another molecule. By contract, an attractive interaction has a negative energy. The dielectric constant is important for the medium.

Hydrogen Bonds

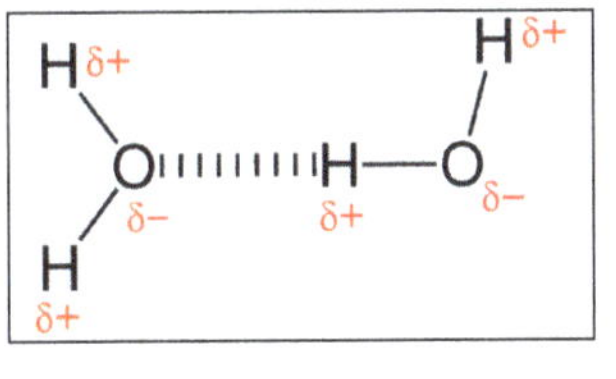

A hydrogen bond is the interaction of a hydrogen atom with an electronegative atom. The electronegative atom can be nitrogen, oxygen, fluorine that comes from another chemical group. Hydrogen bonds are responsible for specific base-pair formation in the DNA double helix. The hydrogen atom in a hydrogen bond is shared by two electronegative atoms such as nitrogen or oxygen. A hydrogen-bond donor is the group that includes an electronegative atom where the hydrogen atom is more tightly bound to and a hydrogen-bond acceptor is when an electronegative atom is less tightly bound to the hydrogen atom. The electronegative atom where the hydrogen atom covalently bonds can pull electron density away from the hydrogen atom creating a positive electronegativity charge. The hydrogen atom can also interact with an atom that has negative electronegativity charge.

Van der Waals Interactions

Van der Waals interaction is the distribution of electronic charge around an atom that fluctuates with time. It is the sum of the attractive or repulsive forces between molecules. The charge distribution is not perfectly symmetric. The attraction increases as two atoms come closer to each other, until they are separated by the van der Waals contact distance. When the distance of the energy is shorter than the van der Waals contact distance, a very strong repulsive force becomes dominant.

Hydrophobic Interactions

Hydrophobic interaction is the physical property of a molecule that is repelled from a mass of water. They are also called hydrophobic exclusions. It is the tendency of hydrocarbons to form intermolecular aggregates in an aqueous medium, and analogous intramolecular interactions.

Ion Induced Dipole

An ion induced dipole is a noncovalent bond interaction that results when the approach of an ion induced a dipole in an atom or in a nonpolar molecule by distributing the arrangement of electrons in the nonpolar species.

Dipole Induced Dipole

A dipole-induced dipole interactions results a nonpolar molecule has its electron density shifted by a charged molecule. If a negatively charged species described by its electron density (for example, in water electron density is centralized on the oxygen atom) is brought near a species with an evenly distributed charged molecule, the electrons go towards the side where the positive species is, and away from the opposite side, creating an artificial dipole.

Chapter 3

Amino Acid, Peptides and Proteins

Amino acids are the building blocks of proteins held together by peptide bonds. The chemical cycle of breaking down proteins into amino acids and simple derivative compounds for the synthesizing of new molecules to be used in the body is called protein breaking. This chapter focuses on proteins and their break down, and protein structures to provide a clearer understanding of the subject matter.

Amino Acid

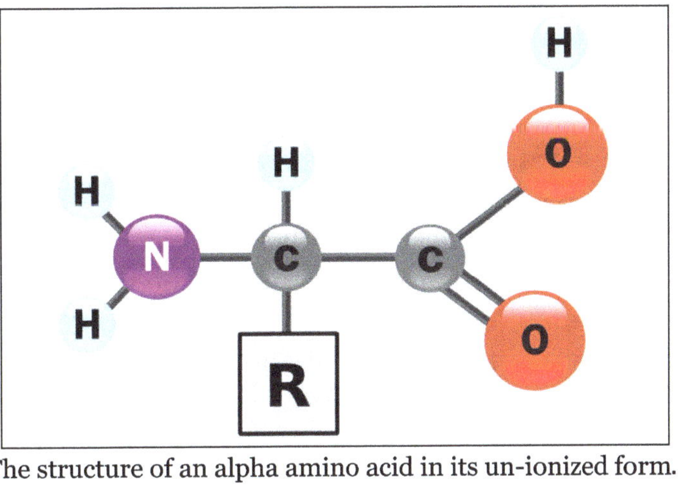

The structure of an alpha amino acid in its un-ionized form.

Amino acids are organic compounds containing amine ($-NH_2$) and carboxyl ($-COOH$) functional groups, along with a side chain (R group) specific to each amino acid. The key elements of an amino acid are carbon, hydrogen, oxygen, and nitrogen, although other elements are found in the side chains of certain amino acids. About 500 amino acids are known (though only 20 appear in the genetic code) and can be classified in many ways. They can be classified according to the core structural functional groups' locations as alpha- (α-), beta- (β-), gamma- (γ-) or delta- (δ-) amino acids; other categories relate to polarity, pH level, and side chain group type (aliphatic, acyclic, aromatic, containing hydroxyl or sulfur, etc.). In the form of proteins, amino acid residues form the second-largest component (water is the largest) of human muscles and other tissues. Beyond their role as residues in proteins, amino acids participate in a number of processes such as neurotransmitter transport and biosynthesis.

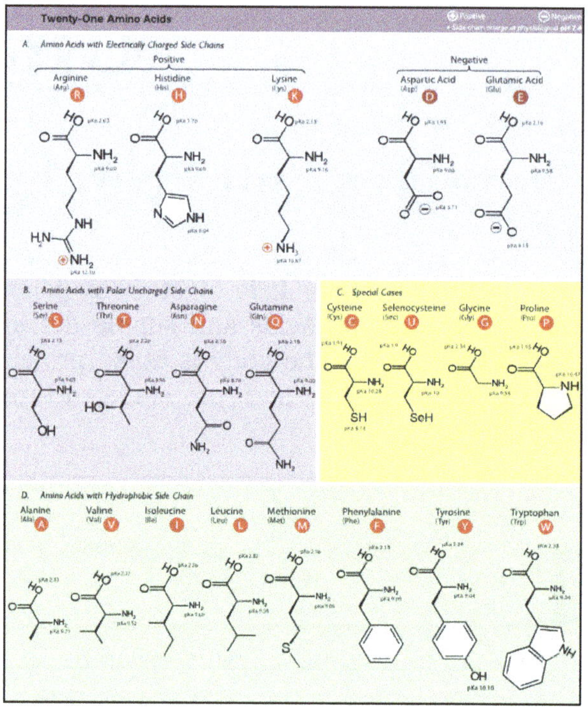

The 21 proteinogenic α-amino acids found in eukaryotes, grouped according to their side chains' pK_a values and charges carried at physiological pH 7.4.

In biochemistry, amino acids having both the amine and the carboxylic acid groups attached to the first (alpha-) carbon atom have particular importance. They are known as 2-, alpha-, or α-amino acids (generic formula $H_2NCHRCOOH$ in most cases, where R is an organic substituent known as a "side chain"); often the term "amino acid" is used to refer specifically to these. They include the 22 proteinogenic ("protein-building") amino acids, which combine into peptide chains ("polypeptides") to form the building-blocks of a vast array of proteins. These are all L-stereoisomers ("left-handed" isomers), although a few D-amino acids ("right-handed") occur in bacterial envelopes, as a neuromodulator (D-serine), and in some antibiotics. Twenty of the proteinogenic amino acids are encoded directly by triplet codons in the genetic code and are known as "standard" amino acids. The other two ("non-standard" or "non-canonical") are selenocysteine (present in many noneukaryotes as well as most eukaryotes, but not coded directly by DNA), and pyrrolysine (found only in some archea and one bacterium). Pyrrolysine and selenocysteine are encoded via variant codons; for example, selenocysteine is encoded by stop codon and SECIS element. *N*-formylmethionine (which is often the initial amino acid of proteins in bacteria, mitochondria, and chloroplasts) is generally considered as a form of methionine rather than as a separate proteinogenic amino acid. Codon–tRNA combinations not found in nature can also be used to "expand" the genetic code and create novel proteins known as alloproteins incorporating non-proteinogenic amino acids.

Many important proteinogenic and non-proteinogenic amino acids have biological functions. For example, in the human brain, glutamate (standard glutamic acid) and gamma-amino-butyric acid ("GABA", non-standard gamma-amino acid) are, respectively, the main excitatory and inhibitory neurotransmitters. Hydroxyproline, a major component of the connective tissue collagen), is synthesised from proline. Glycine is a biosynthetic precursor to porphyrins used in red blood cells. Carnitine is used in lipid transport.

Nine proteinogenic amino acids are called "essential" for humans because they cannot be created from other compounds by the human body and so must be taken in as food. Others may be conditionally essential for certain ages or medical conditions. Essential amino acids may also differ between species.

Because of their biological significance, amino acids are important in nutrition and are commonly used in nutritional supplements, fertilizers, and food technology. Industrial uses include the production of drugs, biodegradable plastics, and chiral catalysts.

History

The first few amino acids were discovered in the early 19th century. In 1806, French chemists Louis-Nicolas Vauquelin and Pierre Jean Robiquet isolated a compound in asparagus that was subsequently named asparagine, the first amino acid to be discovered. Cystine was discovered in 1810, although its monomer, cysteine, remained undiscovered until 1884. Glycine and leucine were discovered in 1820. The last of the 20 common amino acids to be discovered was threonine in 1935 by William Cumming Rose, who also determined the essential amino acids and established the minimum daily requirements of all amino acids for optimal growth.

Usage of the term *amino acid* in the English language is from 1898. Proteins were found to yield amino acids after enzymatic digestion or acid hydrolysis. In 1902, Emil Fischer and Franz Hofmeister proposed that proteins are the result of the formation of bonds between the amino group of one amino acid with the carboxyl group of another, in a linear structure that Fischer termed "peptide".

General Structure

In the structure shown at the top of the page, R represents a side chain specific to each amino acid. The carbon atom next to the carboxyl group (which is therefore numbered 2 in the carbon chain starting from that functional group) is called the α–carbon. Amino acids containing an amino group bonded directly to the alpha carbon are referred to as *alpha amino acids*. These include amino acids such as proline which contain secondary amines, which used to be often referred to as "imino acids".

Isomerism

The alpha amino acids are the most common form found in nature, but only when occurring in the L-isomer. The alpha carbon is a chiral carbon atom, with the exception of glycine which has two indistinguishable hydrogen atoms on the alpha carbon.

Therefore, all alpha amino acids but glycine can exist in either of two enantiomers, called L or D amino acids, which are mirror images of each other. While L-amino acids represent all of the amino acids found in proteins during translation in the ribosome, D-amino acids are found in some proteins produced by enzyme posttranslational modifications after translation and translocation to the endoplasmic reticulum, as in exotic sea-dwelling organisms such as cone snails. They are also abundant components of the peptidoglycan cell walls of bacteria, and D-serine may act as a neurotransmitter in the brain. D-amino acids are used in racemic crystallography to create centrosymmetric crystals, which (depending on the protein) may allow for easier and more robust protein structure determination. The L and D convention for amino acid configuration refers not to the optical activity of the amino acid itself but rather to the optical activity of the isomer of glyceraldehyde from which that amino acid can, in theory, be synthesized (D-glyceraldehyde is dextrorotatory; L-glyceraldehyde is levorotatory). In alternative fashion, the *(S)* and *(R)* designators are used to indicate the absolute stereochemistry. Almost all of the amino acids in proteins are *(S)* at the α carbon, with cysteine being *(R)* and glycine non-chiral. Cysteine has its side chain in the same geometric position as the other amino acids, but the *R/S* terminology is reversed because of the higher atomic number of sulfur compared to the carboxyl oxygen gives the side chain a higher priority, whereas the atoms in most other side chains give them lower priority.

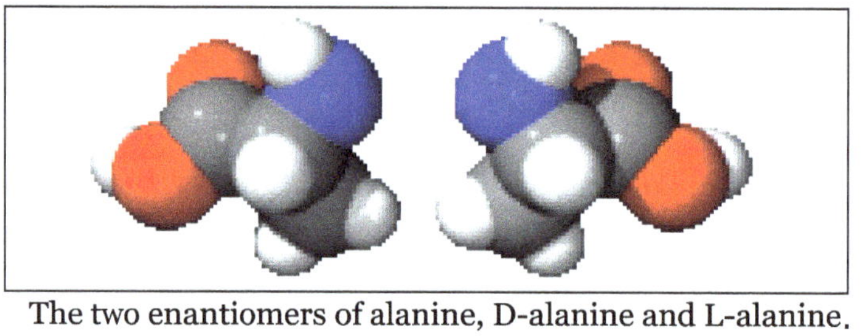

The two enantiomers of alanine, D-alanine and L-alanine.

Side Chains

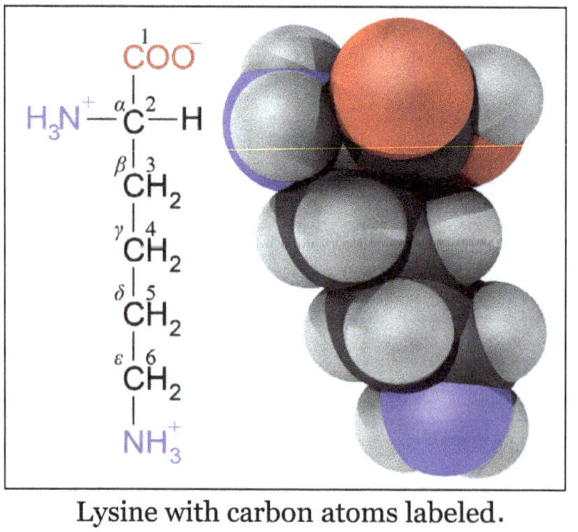

Lysine with carbon atoms labeled.

In amino acids that have a carbon chain attached to the α–carbon the carbons are labeled in order as α, β, γ, δ, and so on. In some amino acids, the amine group is attached to the β or γ-carbon, and these are therefore referred to as *beta* or *gamma amino acids.*

Amino acids are usually classified by the properties of their side chain into four groups. The side chain can make an amino acid a weak acid or a weak base, and a hydrophile if the side chain is polar or a hydrophobe if it is nonpolar. The chemical structures of the 22 standard amino acids, along with their chemical properties.

The phrase "branched-chain amino acids" or BCAA refers to the amino acids having aliphatic side chains that are non-linear; these are leucine, isoleucine, and valine. Proline is the only proteinogenic amino acid whose side-group links to the α-amino group and, thus, is also the only proteinogenic amino acid containing a secondary amine at this position. In chemical terms, proline is, therefore, an imino acid, since it lacks a primary amino group, although it is still classed as an amino acid in the current biochemical nomenclature, and may also be called an "N-alkylated alpha-amino acid".

Zwitterions

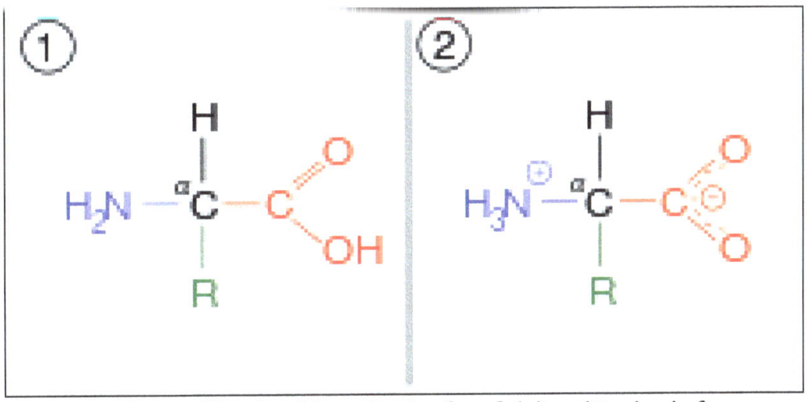

An amino acid in its (1) un-ionized and (2) zwitterionic forms.

The α-carboxylic acid group of amino acids is a weak acid, meaning that it releases a hydron (such as a proton) at moderate pH values. In other words, carboxylic acid groups ($-CO_2H$) can be deprotonated to become negative carboxylates ($-CO_2^-$). The negatively charged carboxylate ion predominates at pH values greater than the pKa of the carboxylic acid group. In a complementary fashion, the α-amine of amino acids is a weak base, meaning that it accepts a proton at moderate pH values. In other words, α-amino groups (NH_2-) can be protonated to become positive α-ammonium groups ($^+NH_3-$). The positively charged α-ammonium group predominates at pH values less than the pKa of the α-ammonium group.

Because all amino acids contain amine and carboxylic acid functional groups, they share amphiprotic properties. Below pH 2.2, the predominant form will have a neutral carboxylic acid group and a positive α-ammonium ion (net charge +1), and above pH 9.4, a negative carboxylate and neutral α-amino group (net charge −1). But at pH between 2.2 and 9.4, an amino acid usually contains both a negative carboxylate and a positive α-ammonium group, as shown in structure (2), so has net zero charge. This molecular state is known as a zwitterion, from the German Zwitter meaning *hermaphrodite* or *hybrid*. The fully neutral form (structure (1)) is a very minor species in aqueous solution throughout the pH range (less than 1 part in 10^7). Amino acids exist as zwitterions also in the solid phase, and crystallize with salt-like properties unlike typical organic acids or amines.

Isoelectric Point

The variation in titration curves when the amino acids are grouped by category can be seen here. With the exception of tyrosine, using titration to differentiate between hydrophobic amino acids is problematic.

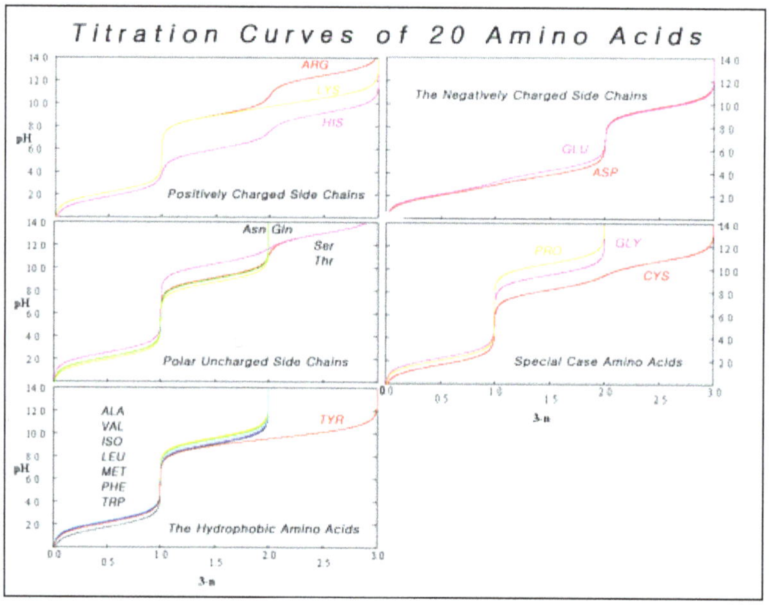

Composite of titration curves of twenty proteinogenic amino acids grouped by side chain category.

At pH values between the two pKa values, the zwitterion predominates, but coexists in dynamic equilibrium with small amounts of net negative and net positive ions. At the exact midpoint between the two pKa values, the trace amount of net negative and trace of net positive ions exactly balance, so that average net charge of all forms present is zero. This pH is known as the isoelectric point pI, so pI = ½(pKa_1 + pKa_2). The individual amino acids all have slightly different pKa values, so have different isoelectric points. For amino acids with charged side chains, the pKa of the side chain is involved. Thus for Asp, Glu with negative side chains, pI = ½(pKa_1 + pKa_R), where pKa_R is the side chain pKa. Cysteine also has potentially negative side chain with pKa_R = 8.14, so pI should be calculated as for Asp and Glu, even though the side chain is not significantly charged at neutral pH. For His, Lys, and Arg with positive side chains, pI = ½(pKa_R

+ pKa_2). Amino acids have zero mobility in electrophoresis at their isoelectric point, although this behaviour is more usually exploited for peptides and proteins than single amino acids. Zwitterions have minimum solubility at their isoelectric point and some amino acids (in particular, with non-polar side chains) can be isolated by precipitation from water by adjusting the pH to the required isoelectric point.

Occurrence and Functions in Biochemistry

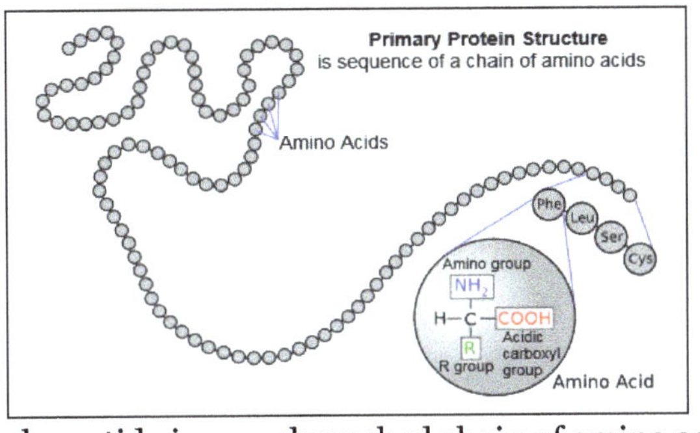

A polypeptide is an unbranched chain of amino acids.

Proteinogenic Amino Acids

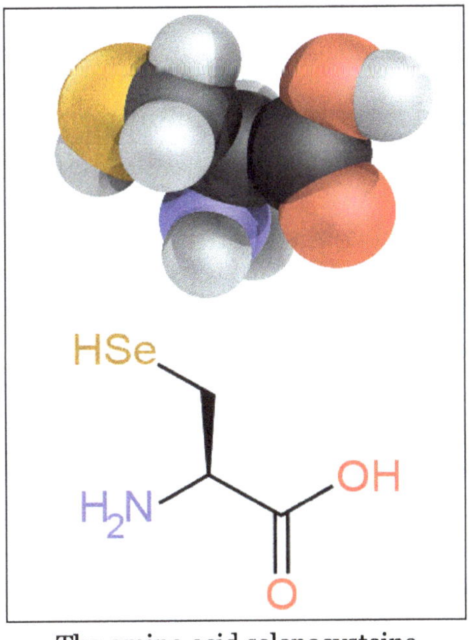

The amino acid selenocysteine.

Amino acids are the structural units (monomers) that make up proteins. They join together to form short polymer chains called peptides or longer chains called either polypeptides or proteins. These polymers are linear and unbranched, with each amino acid within the chain attached to two neighboring amino acids. The process of making proteins is called *translation* and involves the step-by-step addition of amino acids to a growing protein chain by a ribozyme that is called a ribosome. The order in which the amino acids are added is read through the genetic code from an mRNA template, which is a RNA copy of one of the organism's genes.

Twenty-two amino acids are naturally incorporated into polypeptides and are called proteinogenic or natural amino acids. Of these, 20 are encoded by the universal genetic code. The remaining 2, selenocysteine and pyrrolysine, are incorporated into proteins by unique synthetic mechanisms. Selenocysteine is incorporated when the mRNA being translated includes a SECIS element, which causes the UGA codon to encode selenocysteine instead of a stop codon. Pyrrolysine is used by some methanogenic archaea in enzymes that they use to produce methane. It is coded for with the codon UAG, which is normally a stop codon in other organisms. This UAG codon is followed by a PYLIS downstream sequence.

Non-proteinogenic Amino Acids

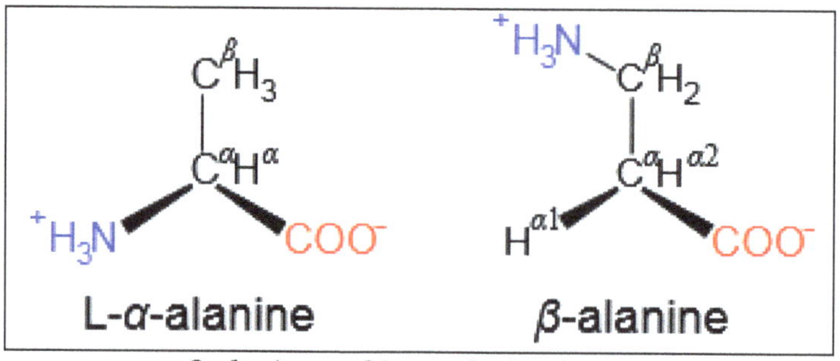

β-alanine and its α-alanine isomer.

Aside from the 22 proteinogenic amino acids, many *non-proteinogenic* amino acids are known. Those either are not found in proteins (for example carnitine, GABA, Levothyroxine) or are not produced directly and in isolation by standard cellular machinery (for example, hydroxyproline and selenomethionine).

Non-proteinogenic amino acids that are found in proteins are formed by post-translational modification, which is modification after translation during protein synthesis. These modifications are often essential for the function or regulation of a protein. Ffor example, the carboxylation of glutamate allows for better binding of calcium cations. connective tissues is composed of hydroxyproline, generated by hydroxylation of proline. Another example is the formation of hypusine in the translation initiation factor EIF5A, through modification of a lysine residue. Such modifications can also determine the localization of the protein, e.g., the addition of long hydrophobic groups can cause a protein to bind to a phospholipid membrane.

Some non-proteinogenic amino acids are not found in proteins. Examples include 2-aminoisobutyric acid and the neurotransmitter gamma-aminobutyric acid. Non-proteinogenic amino acids often occur as intermediates in the metabolic pathways for standard amino acids – for example, ornithine and citrulline occur in the urea cycle, part of amino acid catabolism. A rare exception to the dominance of α-amino acids in biology is the β-amino acid beta alanine (3-aminopropanoic acid), which is used in plants and microorganisms in the synthesis of pantothenic acid (vitamin B_5), a component of coenzyme A.

D-amino Acid Natural Abundance

D-isomers are uncommon in live organisms. For instance, gramicidin is a polypeptide made up from mixture of D- and L-amino acids. Other compounds containing D-amino acid are tyrocidine and valinomycin. These compounds disrupt bacterial cell walls, particularly in Gram-positive bacteria. Only 837 D-amino acids were found in Swiss-Prot database (187 million amino acids analysed).

Non-standard Amino Acids

The 20 amino acids that are encoded directly by the codons of the universal genetic code are called *standard* or *canonical* amino acids. A modified form of methionine (*N*-formylmethionine) is often incorporated in place of methionine as the initial amino acid of proteins in bacteria, mitochondria and chloroplasts. Other amino acids are called *non-standard* or *non-canonical*. Most of the non-standard amino acids are also non-proteinogenic (i.e. they cannot be incorporated into proteins during translation), but two of them are proteinogenic, as they can be incorporated translationally into proteins by exploiting information not encoded in the universal genetic code.

The two non-standard proteinogenic amino acids are selenocysteine (present in many non-eukaryotes as well as most eukaryotes, but not coded directly by DNA) and pyrrolysine (found only in some archaea and one bacterium). The incorporation of these non-standard amino acids is rare. For example, 25 human proteins include selenocysteine (Sec) in their primary structure, and the structurally characterized enzymes (selenoenzymes) employ Sec as the catalytic moiety in their active sites. Pyrrolysine and selenocysteine are encoded via variant codons. For example, selenocysteine is encoded by stop codon and SECIS element.

In Human Nutrition

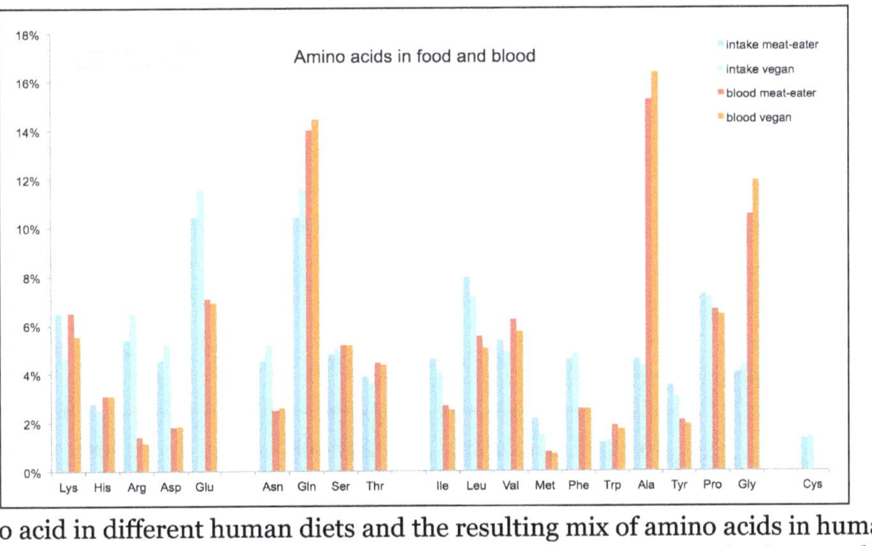

Share of amino acid in different human diets and the resulting mix of amino acids in human blood serum. Glutamate and glutamine are the most frequent in food at over 10%, while alanine, glutamine, and glycine are the most common in blood.

When taken up into the human body from the diet, the 20 standard amino acids either are used to synthesize proteins and other biomolecules or are oxidized to urea and carbon dioxide as a source of energy. The oxidation pathway starts with the removal of the amino group by a transaminase; the amino group is then fed into the urea cycle. The other product of transamidation is a keto acid that enters the citric acid cycle. Glucogenic amino acids can also be converted into glucose, through gluconeogenesis. Of the 20 standard amino acids, nine (His, Ile, Leu, Lys, Met, Phe, Thr, Trp and Val), are called essential amino acids because the human body cannot synthesize them from other compounds at the level needed for normal growth, so they must be obtained from food. In addition, cysteine, taurine, tyrosine, and arginine are considered semiessential amino-acids in children (though taurine is not technically an amino acid), because the metabolic pathways that synthesize these amino acids are not fully developed. The amounts required also depend on the age and health of the individual, so it is hard to make general statements about the dietary requirement for some amino acids. Dietary exposure to the non-standard amino acid BMAA has been linked to human neurodegenerative diseases, including ALS.

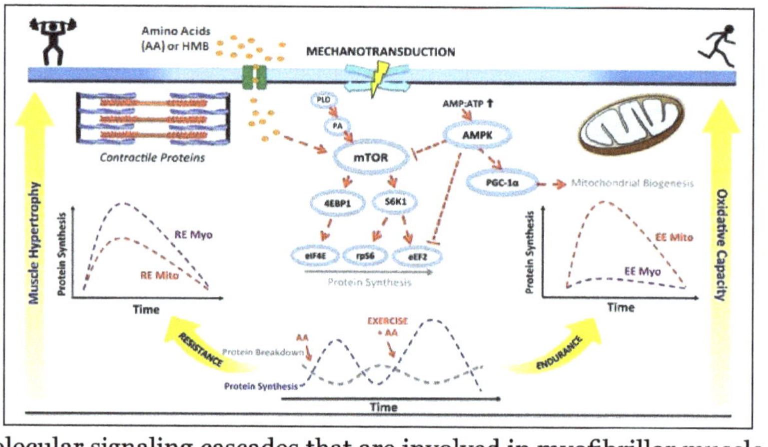

Diagram of the molecular signaling cascades that are involved in myofibrillar muscle protein synthesis and mitochondrial biogenesis in response to physical exercise and specific amino acids or their derivatives (primarily L-leucine and HMB). Many amino acids derived from food protein promote the activation of mTORC1 and increase protein synthesis by signaling through Rag GTPases.

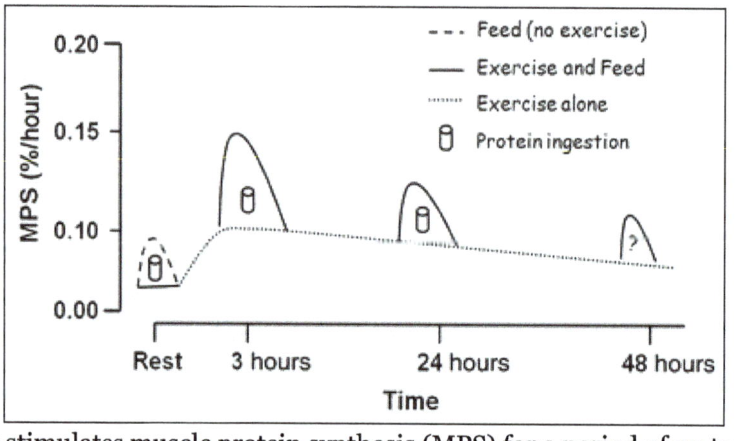

Resistance training stimulates muscle protein synthesis (MPS) for a period of up to 48 hours following exercise (shown by dotted line). Ingestion of a protein-rich meal at any point during this period will augment the exercise-induced increase in muscle protein synthesis (shown by solid lines).

Non-protein Functions

In humans, non-protein amino acids also have important roles as metabolic intermediates, such as in the biosynthesis of the neurotransmitter gamma-amino-butyric acid (GABA). Many amino acids are used to synthesize other molecules, for example:

- Tryptophan is a precursor of the neurotransmitter serotonin.
- Tyrosine (and its precursor phenylalanine) are precursors of the catecholamine neurotransmitters dopamine, epinephrine and norepinephrine and various trace amines.
- Phenylalanine is a precursor of phenethylamine and tyrosine in humans. In plants, it is a precursor of various phenylpropanoids, which are important in plant metabolism.
- Glycine is a precursor of porphyrins such as heme.
- Arginine is a precursor of nitric oxide.
- Ornithine and S-adenosylmethionine are precursors of polyamines.
- Aspartate, glycine, and glutamine are precursors of nucleotides. However, not all of the functions of other abundant non-standard amino acids are known.

Biosynthetic Pathways for Catecholamines and Trace Amines in the Human Brain.

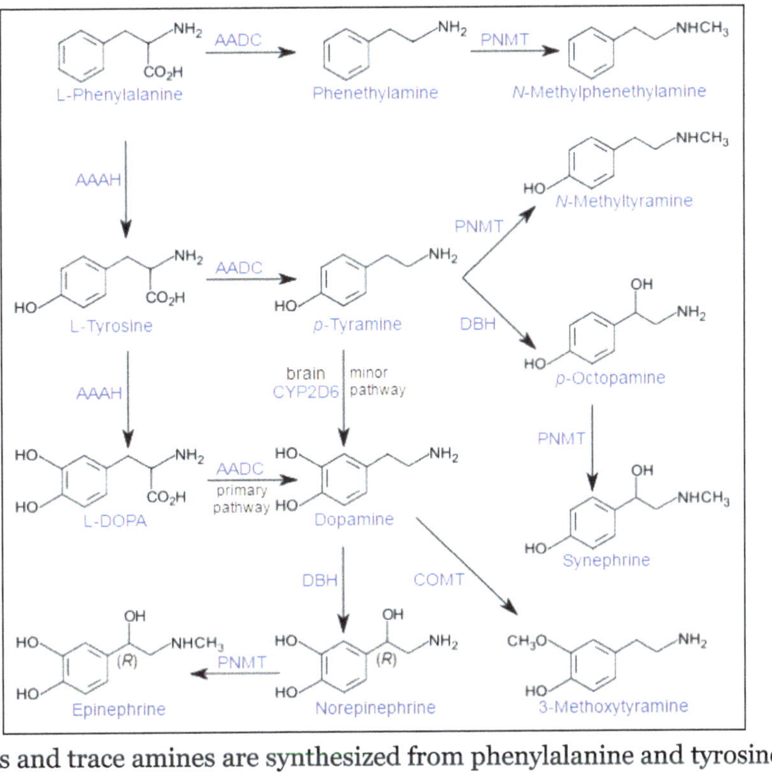

Catecholamines and trace amines are synthesized from phenylalanine and tyrosine in humans.

Some non-standard amino acids are used as defenses against herbivores in plants. For example, canavanine is an analogue of arginine that is found in many legumes, and in particularly large amounts in *Canavalia gladiata* (sword bean). This amino acid protects the plants from predators such as insects and can cause illness in people if some types of legumes are eaten without processing. The non-protein amino acid mimosine is found in other species of legume, in particular *Leucaena leucocephala*. This compound is an analogue of tyrosine and can poison animals that graze on these plants.

Uses in Industry

Amino acids are used for a variety of applications in industry, but their main use is as additives to animal feed. This is necessary, since many of the bulk components of these feeds, such as soybeans, either have low levels or lack some of the essential amino acids: lysine, methionine, threonine, and tryptophan are most important in the production of these feeds. In this industry, amino acids are also used to chelate metal cations in order to improve the absorption of minerals from supplements, which may be required to improve the health or production of these animals.

The food industry is also a major consumer of amino acids, in particular, glutamic acid, which is used as a flavor enhancer, and aspartame (aspartyl-phenylalanine-1-methyl ester) as a low-calorie artificial sweetener. Similar technology to that used for animal nutrition is employed in the human nutrition industry to alleviate symptoms of mineral deficiencies, such as anemia, by improving mineral absorption and reducing negative side effects from inorganic mineral supplementation.

The chelating ability of amino acids has been used in fertilizers for agriculture to facilitate the delivery of minerals to plants in order to correct mineral deficiencies, such as iron chlorosis. These fertilizers are also used to prevent deficiencies from occurring and improving the overall health of the plants. The remaining production of amino acids is used in the synthesis of drugs and cosmetics.

Similarly, some amino acids derivatives are used in pharmaceutical industry. They include 5-HTP (5-hydroxytryptophan) used for experimental treatment of depression, L-DOPA (L-dihydroxyphenylalanine) for Parkinson's treatment, and eflornithine drug that inhibits ornithine decarboxylase and used in the treatment of sleeping sickness.

Expanded Genetic Code

Since 2001, 40 non-natural amino acids have been added into protein by creating a unique codon (recoding) and a corresponding transfer-RNA:aminoacyl – tRNA-synthetase pair to encode it with diverse physicochemical and biological properties in order to be used as a tool to exploring protein structure and function or to create novel or enhanced proteins.

Nullomers

Nullomers are codons that in theory code for an amino acid, however in nature there is a selective bias against using this codon in favor of another, for example bacteria prefer to use CGA instead of AGA to code for arginine. This creates some sequences that do not appear in the genome. This characteristic can be taken advantage of and used to create new selective cancer-fighting drugs and to prevent cross-contamination of DNA samples from crime-scene investigations.

Chemical Building Blocks

Amino acids are important as low-cost feedstocks. These compounds are used in chiral pool synthesis as enantiomerically pure building-blocks.

Amino acids have been investigated as precursors chiral catalysts, e.g., for asymmetric hydrogenation reactions, although no commercial applications exist.

Biodegradable Plastics

Amino acids are under development as components of a range of biodegradable polymers. These materials have applications as environmentally friendly packaging and in medicine in drug delivery and the construction of prosthetic implants. These polymers include polypeptides, polyamides, polyesters, polysulfides, and polyurethanes with amino acids either forming part of their main chains or bonded as side chains. These modifications alter the physical properties and reactivities of the polymers. An interesting example of such materials is polyaspartate, a water-soluble biodegradable polymer that may have applications in disposable diapers and agriculture. Due to its solubility and ability to chelate metal ions, polyaspartate is also being used as a biodegradeable anti-scaling agent and a corrosion inhibitor. In addition, the aromatic amino acid tyrosine is being developed as a possible replacement for toxic phenols such as bisphenol A in the manufacture of polycarbonates.

Reactions

As amino acids have both a primary amine group and a primary carboxyl group, these chemicals can undergo most of the reactions associated with these functional groups. These include nucleophilic addition, amide bond formation, and imine formation for the amine group, and esterification, amide bond formation, and decarboxylation for the carboxylic acid group. The combination of these functional groups allow amino acids to be effective polydentate ligands for metal-amino acid chelates. The multiple side chains of amino acids can also undergo chemical reactions. The types of these reactions are determined by the groups on these side chains and are, therefore, different between the various types of amino acid.

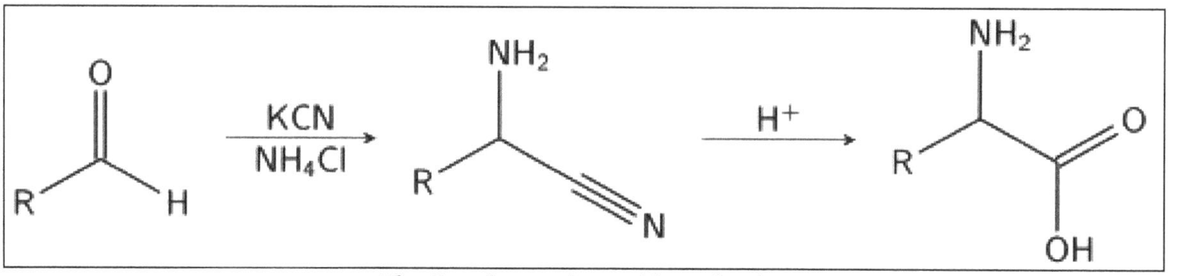

The Strecker amino acid synthesis.

Chemical Synthesis

Several methods exist to synthesize amino acids. One of the oldest methods begins with the bromination at the α-carbon of a carboxylic acid. Nucleophilic substitution with ammonia then converts the alkyl bromide to the amino acid. In alternative fashion, the Strecker amino acid synthesis involves the treatment of an aldehyde with potassium cyanide and ammonia, this produces an α-amino nitrile as an intermediate. Hydrolysis of the nitrile in acid then yields a α-amino acid. Using ammonia or ammonium salts in this reaction gives unsubstituted amino acids, whereas substituting primary and secondary amines will yield substituted amino acids. Likewise, using ketones, instead of aldehydes, gives α,α-disubstituted amino acids. The classical synthesis gives racemic mixtures of α-amino acids as products, but several alternative procedures using asymmetric auxiliaries or asymmetric catalysts have been developed.

At the current time, the most-adopted method is an automated synthesis on a solid support (e.g., polystyrene beads), using protecting groups (e.g., Fmoc and t-Boc) and activating groups (e.g., DCC and DIC).

Peptide Bond Formation

As both the amine and carboxylic acid groups of amino acids can react to form amide bonds, one amino acid molecule can react with another and become joined through an amide linkage. This polymerization of amino acids is what creates proteins. This condensation reaction yields the newly formed peptide bond and a molecule of water. In cells, this reaction does not occur directly; instead, the amino acid is first activated by attachment to a transfer RNA molecule through an ester bond. This aminoacyl-tRNA is produced in an ATP-dependent reaction carried out by an aminoacyl tRNA synthetase. This aminoacyl-tRNA is then a substrate for the ribosome, which catalyzes the attack of the amino group of the elongating protein chain on the ester bond. As a result of this mechanism, all proteins made by ribosomes are synthesized starting at their N-terminus and moving toward their C-terminus.

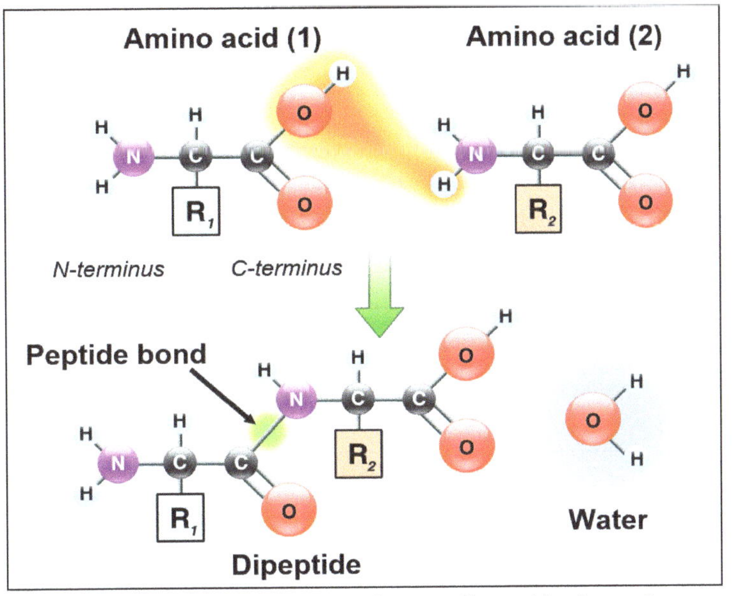

The condensation of two amino acids to form a *dipeptide* through a *peptide bond*.

However, not all peptide bonds are formed in this way. In a few cases, peptides are synthesized by specific enzymes. For example, the tripeptide glutathione is an essential part of the defenses of cells against oxidative stress. This peptide is synthesized in two steps from free amino acids. In the first step, gamma-glutamylcysteine synthetase condenses cysteine and glutamic acid through a peptide bond formed between the side chain carboxyl of the glutamate (the gamma carbon of this side chain) and the amino group of the cysteine. This dipeptide is then condensed with glycine by glutathione synthetase to form glutathione.

In chemistry, peptides are synthesized by a variety of reactions. One of the most-used in solid-phase peptide synthesis uses the aromatic oxime derivatives of amino acids as activated units. These are added in sequence onto the growing peptide chain, which is attached to a solid resin support. The ability to easily synthesize vast numbers of different peptides by varying the types and order of amino acids (using combinatorial chemistry) has made peptide synthesis particularly important in creating libraries of peptides for use in drug discovery through high-throughput screening.

Biosynthesis

In plants, nitrogen is first assimilated into organic compounds in the form of glutamate, formed from alpha-ketoglutarate and ammonia in the mitochondrion. In order to form other amino acids, the plant uses transaminases to move the amino group to another alpha-keto carboxylic acid. For example, aspartate aminotransferase converts glutamate and oxaloacetate to alpha-ketoglutarate and aspartate. Other organisms use transaminases for amino acid synthesis, too.

Nonstandard amino acids are usually formed through modifications to standard amino acids. For example, homocysteine is formed through the transsulfuration pathway or by the demethylation of methionine via the intermediate metabolite S-adenosyl methionine, while hydroxyproline is made by a posttranslational modification of proline.

Microorganisms and plants can synthesize many uncommon amino acids. For example, some microbes make 2-aminoisobutyric acid and lanthionine, which is a sulfide-bridged derivative of alanine. Both of these amino acids are found in peptidic lantibiotics such as alamethicin. However, in plants, 1-aminocyclopropane-1-carboxylic acid is a small disubstituted cyclic amino acid that is a key intermediate in the production of the plant hormone ethylene.

Catabolism

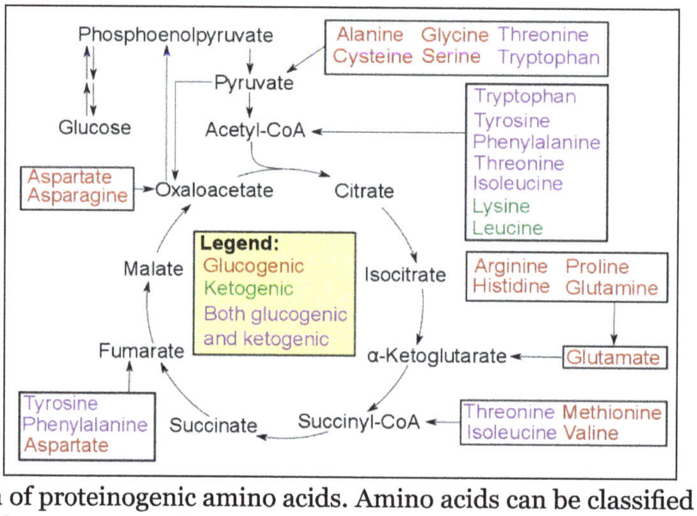

Catabolism of proteinogenic amino acids. Amino acids can be classified according to the properties of their main products as either of the following:
* *Glucogenic*, with the products having the ability to form glucose by gluconeogenesis
* *Ketogenic*, with the products not having the ability to form glucose. These products may still be used for ketogenesis or lipid synthesis.
* Amino acids catabolized into both glucogenic and ketogenic products.

Amino acids must first pass out of organelles and cells into blood circulation via amino acid transporters, since the amine and carboxylic acid groups are typically ionized. Degradation of an amino acid, occurring in the liver and kidneys, often involves deamination by moving its amino group to alpha-ketoglutarate, forming glutamate. This process involves transaminases, often the same as those used in amination during synthesis. In many vertebrates, the amino group is then removed through the urea cycle and is excreted in the form of urea. However, amino acid degradation can produce uric acid or ammonia instead. For example, serine dehydratase converts serine to pyruvate and ammonia. After removal of one or more amino groups, the remainder of the molecule can sometimes be used to synthesize new amino acids, or it can be used for energy by entering glycolysis or the citric acid cycle.

Physicochemical Properties of Amino Acids

The 20 amino acids encoded directly by the genetic code can be divided into several groups based on their properties. Important factors are charge, hydrophilicity or hydrophobicity, size, and functional groups. These properties are important for protein structure and protein–protein interactions. The water-soluble proteins tend to have their hydrophobic residues (Leu, Ile, Val, Phe, and Trp) buried in the middle of the protein, whereas hydrophilic side chains are exposed to the aqueous solvent. (Note that in biochemistry, a residue refers to a specific monomer within the polymeric chain of a polysaccharide, protein or nucleic acid.) The integral membrane proteins tend to have outer rings of exposed hydrophobic amino acids that anchor them into the lipid bilayer. In the case part-way between these two extremes, some peripheral membrane proteins have a patch of hydrophobic amino acids on their surface that locks onto the membrane. In similar fashion, proteins that have to bind to positively charged molecules have surfaces rich with negatively charged amino acids like glutamate and aspartate, while proteins binding to negatively charged molecules have surfaces rich with positively charged chains like lysine and arginine. There are different hydrophobicity scales of amino acid residues.

Some amino acids have special properties such as cysteine, that can form covalent disulfide bonds to other cysteine residues, proline that forms a cycle to the polypeptide backbone, and glycine that is more flexible than other amino acids.

Many proteins undergo a range of posttranslational modifications, when additional chemical groups are attached to the amino acids in proteins. Some modifications can produce hydrophobic lipoproteins, or hydrophilic glycoproteins. These type of modification allow the reversible targeting of a protein to a membrane. For example, the addition and removal of the fatty acid palmitic acid to cysteine residues in some signaling proteins causes the proteins to attach and then detach from cell membranes.

Table of Standard Amino Acid Abbreviations and Properties

Amino Acid	3-Letter	1-Letter	Side chain class	Side chain polarity	Side chain charge (pH 7.4)	Hydropathy index	Absorbance λ_{max}(nm)	ε at λ_{max} (mM^{-1} cm^{-1})	MW (Weight)	Occurrence in proteins (%)
Alanine	Ala	A	aliphatic	nonpolar	neutral	1.8			89.094	8.76
Arginine	Arg	R	basic	basic polar	positive	−4.5			174.203	5.78
Asparagine	Asn	N	amide	polar	neutral	−3.5			132.119	3.93
Aspartic acid	Asp	D	acid	acidic polar	negative	−3.5			133.104	5.49

Cysteine	Cys	C	sulfur-containing	nonpolar	neutral	2.5	250	0.3	121.154	1.38
Glutamic acid	Glu	E	acid	acidic polar	negative	−3.5			147.131	6.32
Glutamine	Gln	Q	amide	polar	neutral	−3.5			146.146	3.9
Glycine	Gly	G	aliphatic	nonpolar	neutral	−0.4			75.067	7.03
Histidine	His	H	basic aromatic	basic polar	positive(10%) neutral(90%)	−3.2	211	5.9	155.156	2.26
Isoleucine	Ile	I	aliphatic	nonpolar	neutral	4.5			131.175	5.49
Leucine	Leu	L	aliphatic	nonpolar	neutral	3.8			131.175	9.68
Lysine	Lys	K	basic	basic polar	positive	−3.9			146.189	5.19
Methionine	Met	M	sulfur-containing	nonpolar	neutral	1.9			149.208	2.32
Phenylalanine	Phe	F	aromatic	nonpolar	neutral	2.8	257, 206, 188	0.2, 9.3, 60.0	165.192	3.87
Proline	Pro	P	cyclic	nonpolar	neutral	−1.6			115.132	5.02
Serine	Ser	S	hydroxyl-containing	polar	neutral	−0.8			105.093	7.14
Threonine	Thr	T	hydroxyl-containing	polar	neutral	−0.7			119.119	5.53
Tryptophan	Trp	W	aromatic	nonpolar	neutral	−0.9	280, 219	5.6, 47.0	204.228	1.25
Tyrosine	Tyr	Y	aromatic	polar	neutral	−1.3	274, 222, 193	1.4, 8.0, 48.0	181.191	2.91
Valine	Val	V	aliphatic	nonpolar	neutral	4.2			117.148	6.73

Two additional amino acids are in some species coded for by codons that are usually interpreted as stop codons.

21st and 22nd amino acids	3-Letter	1-Letter	MW(Weight)
Selenocysteine	Sec	U	168.064
Pyrrolysine	Pyl	O	255.313

In addition to the specific amino acid codes, placeholders are used in cases where chemical or crystallographic analysis of a peptide or protein cannot conclusively determine the identity of a residue. They are also used to summarise conserved protein sequence motifs. The use of single letters to indicate sets of similar residues is similar to the use of abbreviation codes for degenerate bases.

Ambiguous Amino Acids	3-Letter	1-Letter	
Any / unknown	Xaa	X	All
Asparagine or aspartic acid	Asx	B	D, N
Glutamine or glutamic acid	Glx	Z	E, Q
Leucine or Isoleucine	Xle	J	I, L
Hydrophobic		Φ	V, I, L, F, W, Y, M
Aromatic		Ω	F, W, Y, H
Aliphatic (Non-Aromatic)		Ψ	V, I, L, M
Small		π	P, G, A, S
Hydrophilic		ζ	S, T, H, N, Q, E, D, K, R
Positively charged		+	K, R, H
Negatively charged		−	D, E

Unk is sometimes used instead of Xaa, but is less standard.

In addition, many non-standard amino acids have a specific code. For example, several peptide drugs, such as Bortezomib and MG132, are artificially synthesized and retain their protecting groups, which have specific codes. Bortezomib is Pyz-Phe-boroLeu, and MG132 is Z-Leu-Leu-Leu-al. To aid in the analysis of protein structure, photo-reactive amino acid analogs are available. These include photoleucine (pLeu) and photomethionine (pMet).

Peptide Bond

The covalent bond that holds two adjacent amino acid residues together is known as a peptide bond, which is formed between the carboxyl group of one amino acid and the amino group of other amino acid and is accompanied by release of a water molecule. The peptide bond is stabilized by resonance structure. The amide bond exhibits partial double bond character and is planar. In other words, it can exist in "cis" and "trans" form. In the unfolded form of a given protein, the peptide bonds have the liberty to take up either of the two forms; however the folded conformation has the peptide bond in a single form alone. The "trans" form is usually preferred as it's conformation is stable as compared to the "cis" form (Exception: Proline, which can exist in "cis" as well as "trans" form). The psi and phi are the angles of rotation about the bond between the a-carbon atom and carboxyl and amino groups, respectively. These angles determine which protein conformations will be favourable during protein folding.

Protein Structure

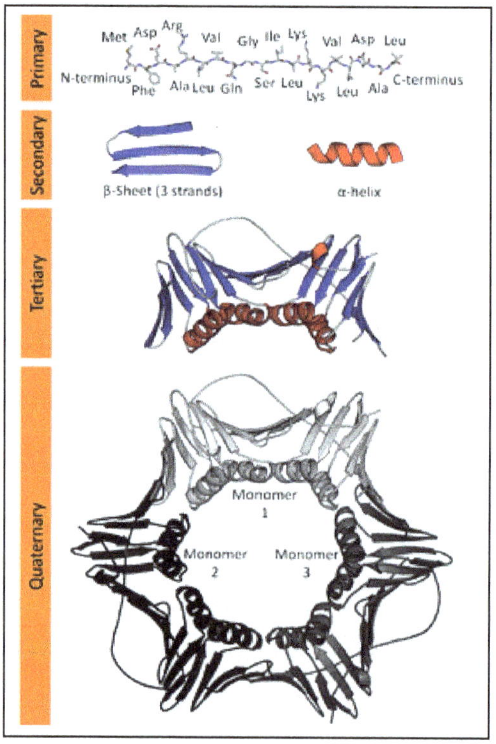

Protein structure, using PCNA as an example. (PDB: 1AXC)

Protein structure is the three-dimensional arrangement of atoms in a protein molecule. Proteins are polymers — specifically polypeptides — formed from sequences of amino acids, the monomers of the polymer. A single amino acid monomer may also be called a residue indicating a repeating unit of a polymer. Proteins form by amino acids undergoing condensation reactions, in which the amino acids lose one water molecule per reaction in order to attach to one another with a peptide bond. By convention, a chain under 30 amino acids is often identified as a peptide, rather than a protein. To be able to perform their biological function, proteins fold into one or more specific spatial conformations driven by a number of non-covalent interactions such as hydrogen bonding, ionic interactions, Van der Waals forces, and hydrophobic packing. To understand the functions of proteins at a molecular level, it is often necessary to determine their three-dimensional structure. This is the topic of the scientific field of structural biology, which employs techniques such as X-ray crystallography, NMR spectroscopy, and dual polarisation interferometry to determine the structure of proteins.

Protein structures range in size from tens to several thousand amino acids. By physical size, proteins are classified as nanoparticles, between 1–100 nm. Very large aggregates can be formed from protein subunits. For example, many thousands of actin molecules assemble into a microfilament.

A protein may undergo reversible structural changes in performing its biological function. The alternative structures of the same protein are referred to as different conformational isomers, or simply, conformations, and transitions between them are called conformational changes.

Levels of Protein Structure

There are four distinct levels of protein structure.

Amino Acid Residues

Each α-amino acid consists of a backbone that is present in all the amino acid types and a side chain that is unique to each type of residue. An exception from this rule is proline. Because the carbon atom is bound to four different groups it is chiral, however only one of the isomers occur in biological proteins. Glycine however, is not chiral since its side chain is a hydrogen atom. A simple mnemonic for correct L-form is "CORN": when the C_α atom is viewed with the H in front, the residues read "CO-R-N" in a clockwise direction.

Primary Structure

The primary structure of a protein refers to the linear sequence of amino acids in the polypeptide chain. The primary structure is held together by covalent bonds such as peptide bonds, which are made during the process of protein biosynthesis. The two ends of the polypeptide chain are referred to as the carboxyl terminus (C-terminus) and the amino terminus (N-terminus) based on the nature of the free group on each extremity. Counting of residues always starts at the N-terminal end (NH_2-group), which is the end where the amino group is not involved in a peptide bond. The primary structure of a protein is determined by the gene corresponding to the protein. A specific sequence of nucleotides in DNA is transcribed into mRNA, which is read by the ribosome in a process called translation. The sequence of amino acids in insulin was discovered by Frederick Sanger, establishing that proteins have defining amino acid sequences. The sequence of a protein is unique to that protein, and defines the structure and function of the protein. The sequence of a protein can be determined by methods such as Edman degradation or tandem mass spectrometry. Often, however, it is read directly from the sequence of the gene using the genetic code. It is strictly recommended to use the words "amino acid residues" when discussing proteins because when a peptide bond is formed, a water molecule is lost, and therefore proteins are made up of amino acid residues. Post-translational modification such as disulfide bond formation, phosphorylations and glycosylations are usually also considered a part of the primary structure, and cannot be read from the gene. For example, insulin is composed of 51 amino acids in 2 chains. One chain has 31 amino acids, and the other has 20 amino acids.

Secondary Structure

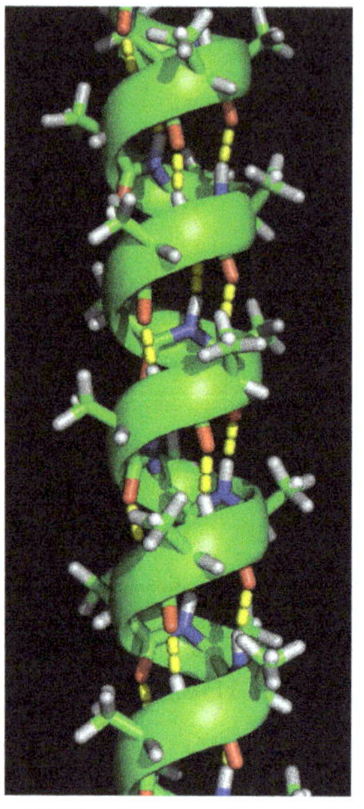

An α-helix with hydrogen bonds (yellow dots).

Secondary structure refers to highly regular local sub-structures on the actual polypeptide backbone chain. Two main types of secondary structure, the α-helix and the β-strand or β-sheets, were suggested in 1951 by Linus Pauling and coworkers. These secondary structures are defined by patterns of hydrogen bonds between the main-chain peptide groups. They have a regular geometry, being constrained to specific values of the dihedral angles ψ and φ on the Ramachandran plot. Both the α-helix and the β-sheet represent a way of saturating all the hydrogen bond donors and acceptors in the peptide backbone. Some parts of the protein are ordered but do not form any regular structures. They should not be confused with random coil, an unfolded polypeptide chain lacking any fixed three-dimensional structure. Several sequential secondary structures may form a "supersecondary unit".

Tertiary Structure

Tertiary structure refers to the three-dimensional structure of monomeric and multimeric protein molecules. The α-helixes and β-pleated-sheets are folded into a compact globular structure. The folding is driven by the *non-specific* hydrophobic interactions, the burial of hydrophobic residues from water, but the structure is stable only when the parts of a protein domain are locked into place by *specific* tertiary interactions, such as salt bridges, hydrogen bonds, and the tight packing of side chains and disulfide bonds. The disulfide bonds are extremely rare in cytosolic proteins, since the cytosol (intracellular fluid) is generally a reducing environment.

Quaternary Structure

Quaternary structure is the three-dimensional structure of a multi-subunit protein and how the subunits fit together. In this context, the quaternary structure is stabilized by the same non-covalent interactions and disulfide bonds as the tertiary structure. Complexes of two or more polypeptides (i.e. multiple subunits) are called multimers. Specifically it would be called a dimer if it contains two subunits, a trimer if it contains three subunits, a tetramer if it contains four subunits, and a pentamer if it contains five subunits. The subunits are frequently related to one another by symmetry operations, such as a 2-fold axis in a dimer. Multimers made up of identical subunits are referred to with a prefix of "homo-" (e.g. a homotetramer) and those made up of different subunits are referred to with a prefix of "hetero-", for example, a heterotetramer, such as the two alpha and two beta chains of hemoglobin.

Domains, Motifs, and Folds in Protein Structure

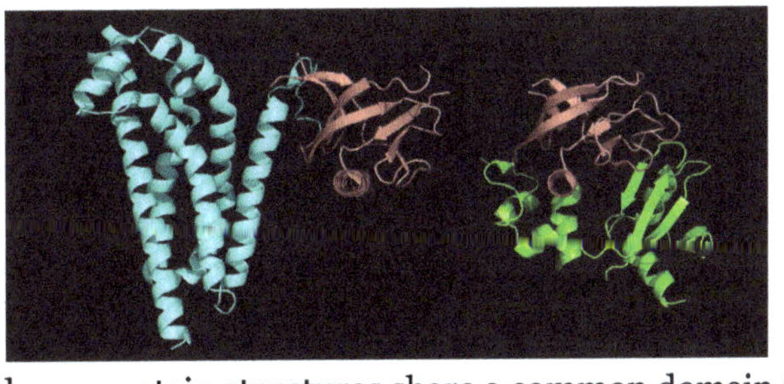

Protein domains. The two shown protein structures share a common domain (maroon), the PH domain, which is involved in phosphatidylinositol (3,4,5)-trisphosphate binding.

Proteins are frequently described as consisting of several structural units. These units include domains, motifs, and folds. Despite the fact that there are about 100,000 different proteins expressed in eukaryotic systems, there are many fewer different domains, structural motifs and folds.

Structural Domain

A structural domain is an element of the protein's overall structure that is self-stabilizing and often folds independently of the rest of the protein chain. Many domains are not unique to the protein products of one gene or one gene family but instead appear in a variety of proteins. Domains often are named and singled out because they figure prominently in the biological function of the protein they belong to; for example, the "calcium-binding domain of calmodulin". Because they are independently stable, domains can be "swapped" by genetic engineering between one protein and another to make chimera proteins.

Structural and Sequence Motif

The structural and sequence motifs refer to short segments of protein three-dimensional structure or amino acid sequence that were found in a large number of different proteins.

Supersecondary Structure

The supersecondary structure refers to a specific combination of secondary structure elements, such as β-α-β units or a helix-turn-helix motif. Some of them may be also referred to as structural motifs.

Protein Fold

A protein fold refers to the general protein architecture, like a helix bundle, β-barrel, Rossman fold or different "folds" provided in the Structural Classification of Proteins database. A related concept is protein topology that refers to the arrangement of contacts within the protein.

Superdomain

A superdomain consists of two or more nominally unrelated structural domains that are inherited as a single unit and occur in different proteins. An example is provided by the protein tyrosine phosphatase domain and C2 domain pair in PTEN, several tensin proteins, auxilin and proteins in plants and fungi. The PTP-C2 superdomain evidently came into existence prior to the divergence of fungi, plants and animals is therefore likely to be about 1.5 billion years old.

Protein Folding

Once translated by a ribosome, each polypeptide folds into its characteristic three-dimensional structure from a random coil. Since the fold is maintained by a network of interactions between amino acids in the polypeptide, the native state of the protein chain is determined by the amino acid sequence (Anfinsen's dogma).

Protein Structure Determination

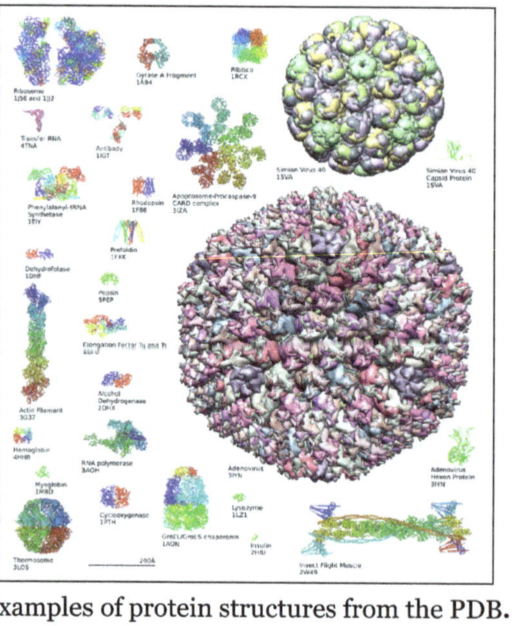

Examples of protein structures from the PDB.

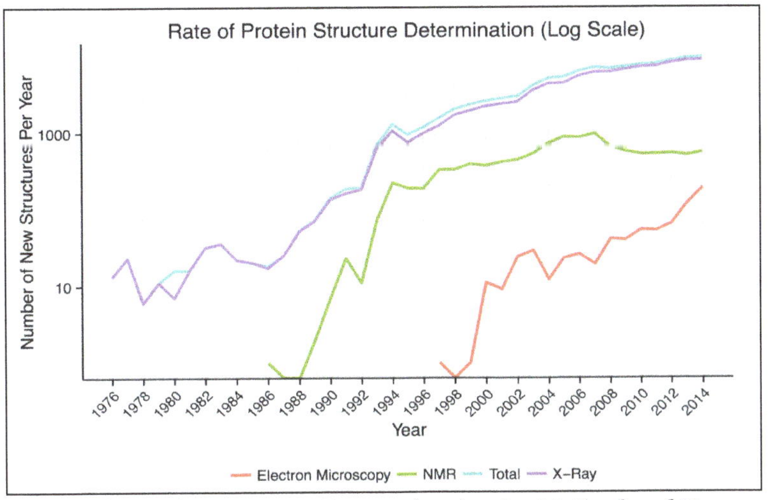

Rate of Protein Structure Determination by Method and Year.

Around 90% of the protein structures available in the Protein Data Bank have been determined by X-ray crystallography. This method allows one to measure the three-dimensional (3-D) density distribution of electrons in the protein, in the crystallized state, and thereby infer the 3-D coordinates of all the atoms to be determined to a certain resolution. Roughly 9% of the known protein structures have been obtained by nuclear magnetic resonance techniques. The secondary structure composition can be determined via circular dichroism. Vibrational spectroscopy can also be used to characterize the conformation of peptides, polypeptides, and proteins. Two-dimensional infrared spectroscopy has become a valuable method to investigate the structures of flexible peptides and proteins that cannot be studied with other methods. Cryo-electron microscopy has recently become a means of determining protein structures to high resolution, less than 5 ångströms or 0.5 nanometer, and is anticipated to increase in power as a tool for high resolution work in the next decade. This technique is still a valuable resource for researchers working with very large protein complexes such as virus coat proteins and amyloid fibers. A more qualitative picture of protein structure is often obtained by proteolysis, which is also useful to screen for more crystallizable protein samples. Novel implementations of this approach, including fast parallel proteolysis (FASTpp), can probe the structured fraction and its stability without the need for purification.

Protein Sequence Analysis: Ensembles

Proteins are often thought of as relatively stable structures that have a set tertiary structure and experience conformational changes as a result of being modified by other proteins or as part of enzymatic activity. However proteins have varying degrees of stability and some of the less stable variants are intrinsically disordered proteins. These proteins exist and function in a relatively 'disordered' state lacking a stable tertiary structure. As a result, they are difficult to describe in a standard protein structure model that was designed for proteins with a fixed tertiary structure. Conformational ensembles have been devised as a way to provide a more accurate and 'dynamic' representation of the conformational state of intrinsically disordered proteins. Conformational ensembles function by attempting to represent the various conformations of intrinsically disordered proteins within an ensemble file (the type found at the Protein Ensemble Database).

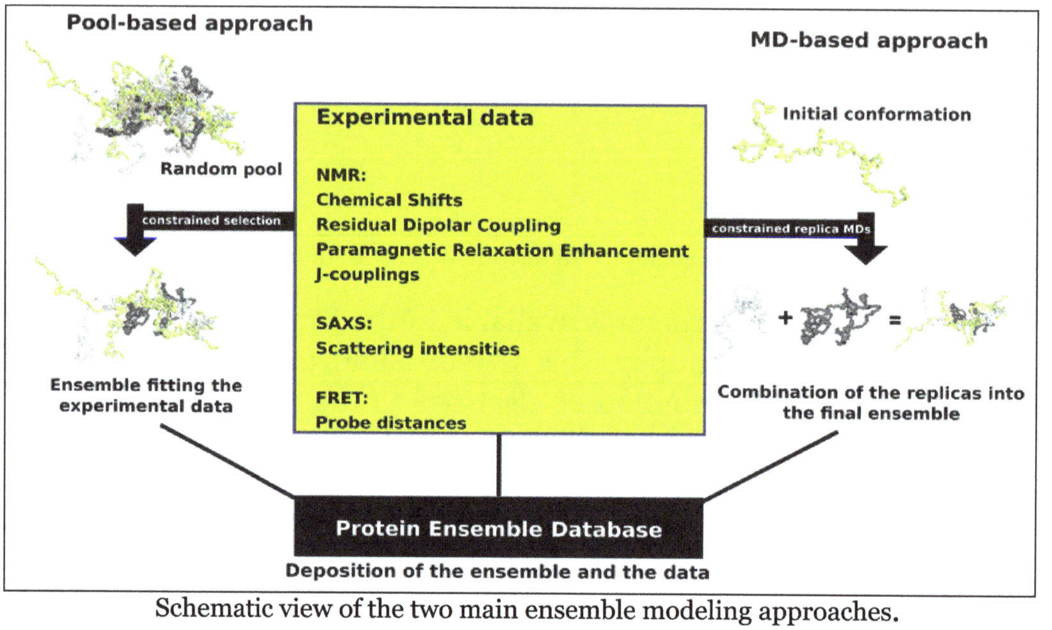

Schematic view of the two main ensemble modeling approaches.

Protein ensemble files are a representation of a protein that can be considered to have a flexible structure. Creating these files requires determining which of the various theoretically possible protein conformations actually exist. One approach is to apply computational algorithms to the protein data in order to try to determine the most likely set of conformations for an ensemble file.

There are multiple methods for preparing data for the Protein Ensemble Database that fall into two general methodologies – pool and molecular dynamics (MD) approaches (diagrammed in the figure). The pool based approach uses the protein's amino acid sequence to create a massive pool of random conformations. This pool is then subjected to more computational processing that creates a set of theoretical parameters for each conformation based on the structure. Conformational subsets from this pool whose average theoretical parameters closely match known experimental data for this protein are selected.

The molecular dynamics approach takes multiple random conformations at a time and subjects all of them to experimental data. Here the experimental data is serving as limitations to be placed on the conformations (e.g. known distances between atoms). Only conformations that manage to remain within the limits set by the experimental data are accepted. This approach often applies large amounts of experimental data to the conformations which is a very computationally demanding task.

Protein	Data Type	Protocol	PED ID
Sic1/Cdc4	NMR and SAXS	Pool-based	PED9AAA
p15 PAF	NMR and SAXS	Pool-based	PED6AAA
MKK7	NMR	Pool-based	PED5AAB
Beta-synuclein	NMR	MD-based	PED1AAD
P27 KID	NMR	MD-based	PED2AAA

Structure Classification

Protein structures can be grouped based on their similarity or a common evolutionary origin. The Structural Classification of Proteins database and CATH database provide two different structural classifications of proteins. Shared structure between proteins is considered evidence of evolutionary relatedness between proteins and is used group proteins together into protein superfamilies.

Computational Prediction of Protein Structure

The generation of a protein sequence is much easier than the determination of a protein structure. However, the structure of a protein gives much more insight in the function of the protein than its sequence. Therefore, a number of methods for the computational prediction of protein structure from its sequence have been developed. *Ab initio* prediction methods use just the sequence of the protein. Threading and homology modeling methods can build a 3-D model for a protein of unknown structure from experimental structures of evolutionarily-related proteins, called a protein family.

Protein Folding

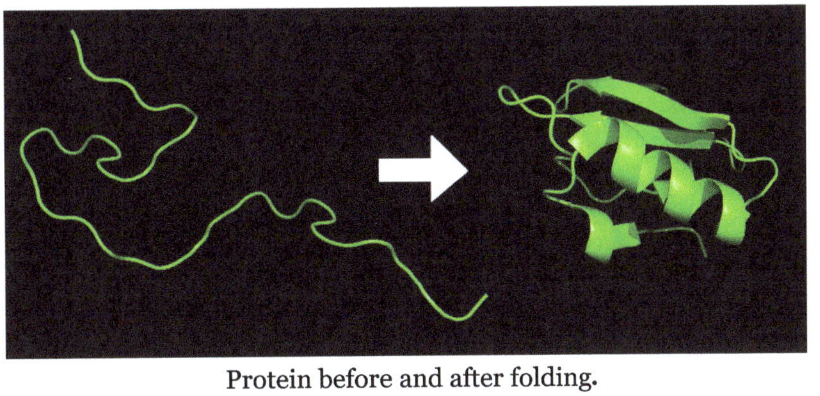

Protein before and after folding.

Protein folding is the physical process by which a protein chain acquires its native 3-dimensional structure, a conformation that is usually biologically functional, in an expeditious and reproducible manner. It is the physical process by which a polypeptide folds into its characteristic and functional three-dimensional structure from random coil. Each protein exists as an unfolded polypeptide or random coil when translated from a sequence of mRNA to a linear chain of amino acids. This polypeptide lacks any stable (long-lasting) three-dimensional structure (the left hand side of the first figure). As the polypeptide chain is being synthesized by the ribosome, the linear chain begins to fold into its three dimensional structure. Folding begins to occur even during translation of the polypeptide chain. Amino acids interact with each other to produce a well-defined three-dimensional structure, the folded protein, known as the native state. The resulting three-dimensional structure is determined by the amino acid sequence or primary structure (Anfinsen's dogma). The energy landscape describes the folding pathways in which the unfolded protein is able to assume its native state. Experiments beginning in the 1980s indicate the codon for an amino acid can also influence protein structure.

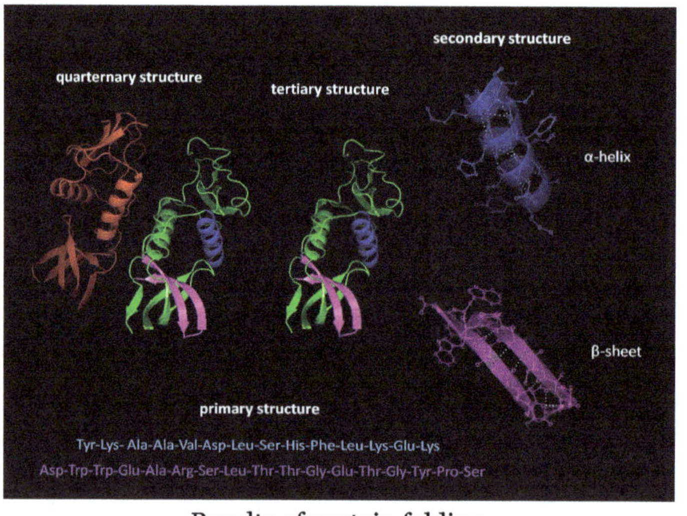

Results of protein folding.

The correct three-dimensional structure is essential to function, although some parts of functional proteins may remain unfolded, so that protein dynamics is important. Failure to fold into native structure generally produces inactive proteins, but in some instances misfolded proteins have modified or toxic functionality. Several neurodegenerative and other diseases are believed to result from the accumulation of amyloid fibrils formed by *misfolded* proteins. Many allergies are caused by incorrect folding of some proteins, because the immune system does not produce antibodies for certain protein structures.

Process of Protein Folding

Primary Structure

The primary structure of a protein, its linear amino-acid sequence, determines its native conformation. The specific amino acid residues and their position in the polypeptide chain are the determining factors for which portions of the protein fold closely together and form its three dimensional conformation. The amino acid composition is not as important as the sequence. The essential fact of folding, however, remains that the amino acid sequence of each protein contains the information that specifies both the native structure and the pathway to attain that state. This is not to say that nearly identical amino acid sequences always fold similarly. Conformations differ based on environmental factors as well; similar proteins fold differently based on where they are found.

Secondary Structure

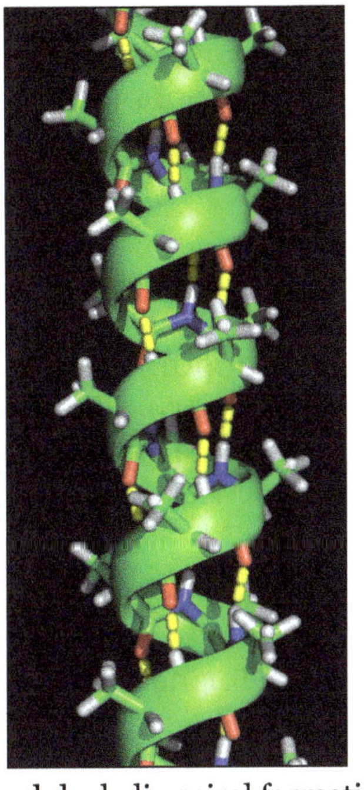

The alpha helix spiral formation.

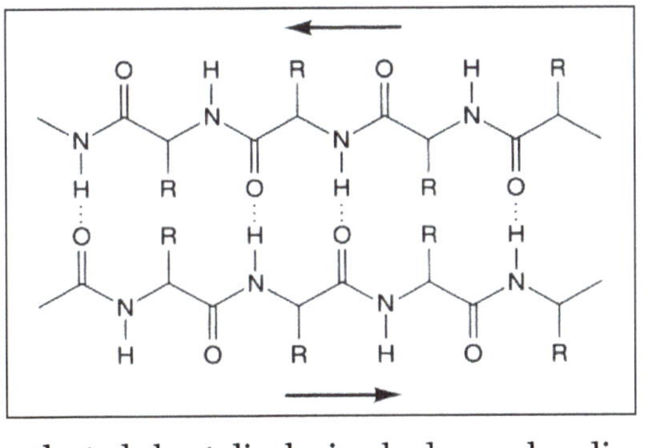

An anti-parallel beta pleated sheet displaying hydrogen bonding within the backbone.

Formation of a secondary structure is the first step in the folding process that a protein takes to assume its native structure. Characteristic of secondary structure are the structures known as alpha helices and beta sheets that fold rapidly because they are stabilized by intramolecular hydrogen bonds, as was first characterized by Linus Pauling. Formation of intramolecular hydrogen bonds provides another important contribution to protein stability. Alpha helices are formed by hydrogen bonding of the backbone to form a spiral shape. The beta pleated sheet is a structure that forms with the backbone bending over itself to form the hydrogen bonds. The hydrogen bonds are between the amide hydrogen and carbonyl carbon of the peptide bond. There exists anti-parallel beta pleated sheets and parallel beta pleated sheets where the stability of the hydrogen bonds is stronger in the anti-parallel beta sheet as it hydrogen bonds with the ideal 180 degree angle compared to the slanted hydrogen bonds formed by parallel sheets. The alpha helices and beta pleated sheets can be amphipathic in nature, or contain a hydrophilic portion and a hydrophobic portion. This property of these secondary structures aids in the folding of the protein as it aligns the helices and sheets in such a way where the hydrophilic sides are facing the aqueous environment surrounding the protein and the hydrophobic sides are facing the hydrophobic core of the protein. Secondary structure hierarchically gives way to tertiary structure formation.

Tertiary Structure

Once the protein's tertiary structure is formed and stabilized by the hydrophobic interactions, there may also be covalent bonding in the form of disulfide bridges formed between two cysteine residues. Tertiary structure of a protein involves a single polypeptide chain; however, additional interactions of folded polypeptide chains give rise to quaternary structure formation.

Quaternary Structure

Tertiary structure may give way to the formation of quaternary structure in some proteins, which usually involves the "assembly" or "coassembly" of subunits that have already folded; in other words, multiple polypeptide chains could interact to form a fully functional quaternary protein.

Driving Force of Protein Folding

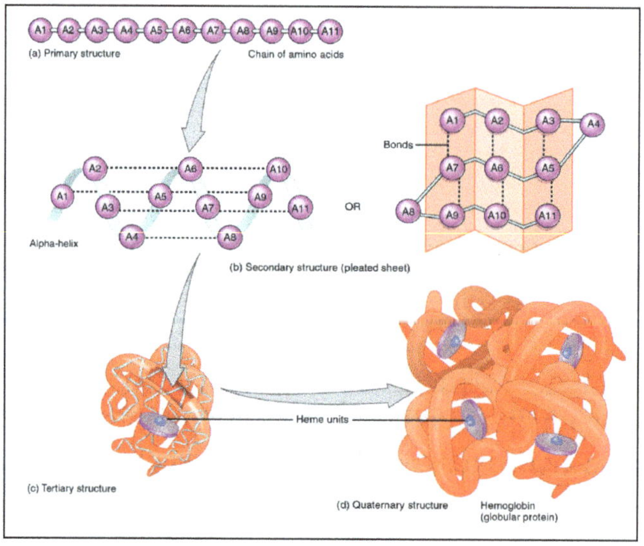

All forms of protein structure summarized.

Folding is a spontaneous process that is mainly guided by hydrophobic interactions, formation of intramolecular hydrogen bonds, van der Waals forces, and it is opposed by conformational entropy. The process of folding often begins co-translationally, so that the N-terminus of the protein begins to fold while the C terminal portion of the protein is still being synthesized by the ribosome; however, a protein molecule may fold spontaneously during or after biosynthesis. While these macromolecules may be regarded as "folding themselves", the process also depends on the solvent (water or lipid bilayer), the concentration of salts, the pH, the temperature, the possible presence of cofactors and of molecular chaperones. Proteins will have limitations on their folding abilities by the restricted bending angles or conformations that are possible. These allowable angles of protein folding are described with a two-dimensional plot known as the Ramachandran plot, depicted with psi and phi angles of allowable rotation.

Hydrophobic Effect

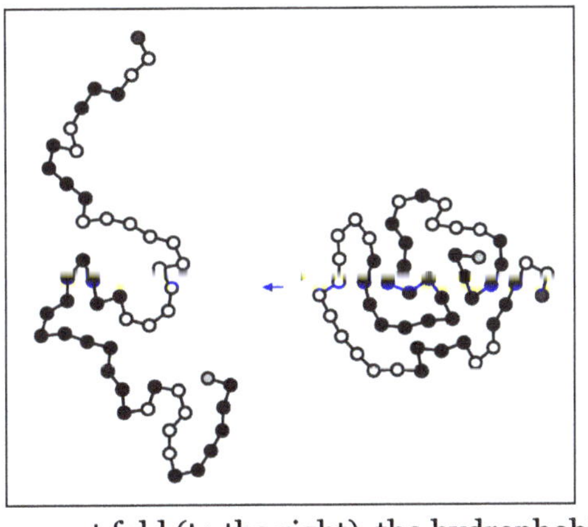

Hydrophobic collapse. In the compact fold (to the right), the hydrophobic amino acids (shown as black spheres) collapse toward the center to become shielded from aqueous environment.

Protein folding must be thermodynamically favorable within a cell in order for it to be a spontaneous reaction. Since it is known that protein folding is a spontaneous reaction, then it must assume a negative Gibbs free energy value. Gibbs free energy in protein folding is directly related to enthalpy and entropy. For a negative delta G to arise and for protein folding to become thermodynamically favorable, then either enthalpy, entropy, or both terms must be favorable.

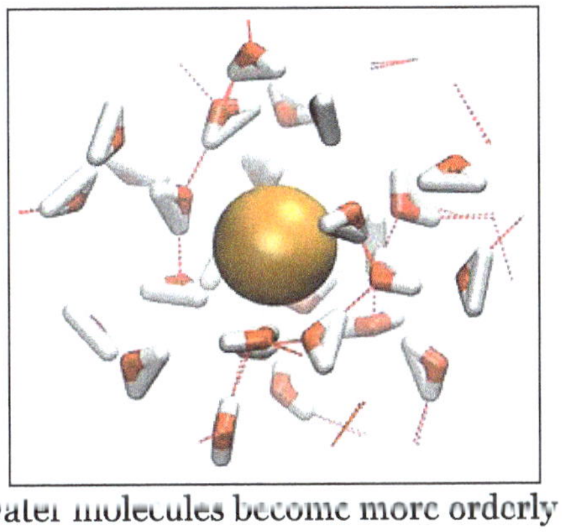

Entropy is decreased as the water molecules become more orderly near the hydrophobic solute.

Minimizing the number of hydrophobic side-chains exposed to water is an important driving force behind the folding process. The hydrophobic effect is the phenomenon in which the hydrophobic chains of a protein collapse into the core of the protein (away from the hydrophilic environment). In an aqueous environment, the water molecules tend to aggregate around the hydrophobic regions or side chains of the protein, creating water shells of ordered water molecules. An ordering of water molecules around a hydrophobic region increases order in a system and therefore contributes a negative change in entropy (less entropy in the system). The water molecules are fixed in these water cages which drives the hydrophobic collapse, or the inward folding of the hydrophobic groups. The hydrophobic collapse introduces entropy back to the system via the breaking of the water cages which frees the ordered water molecules. The multitude of hydrophobic groups interacting within the core of the globular folded protein contributes a significant amount to protein stability after folding, because of the vastly accumulated van der Waals forces (specifically London Dispersion forces). The hydrophobic effect exists as a driving force in thermodynamics only if there is the presence of an aqueous medium with an amphiphilic molecule containing a large hydrophobic region. The strength of hydrogen bonds depends on their environment; thus, H-bonds enveloped in a hydrophobic core contribute more than H-bonds exposed to the aqueous environment to the stability of the native state.

Chaperones

Chaperones are a class of proteins that aid in the correct folding of other proteins *in vivo*. Chaperones exist in all cellular compartments and interact with the polypeptide chain in order to allow the native three-dimensional conformation of the protein to form; however, chaperones themselves are not included in the final structure of the protein they are assisting in. Chaperones may assist in folding even when the nascent polypeptide is being synthesized by the ribosome. Molecular chaperones operate by binding to stabilize an otherwise unstable structure of a protein in its folding pathway, but chaperones do not contain the necessary information to know the correct native structure of the protein they are aiding; rather, chaperones work by preventing incorrect folding conformations. In this way, chaperones do not actually increase the rate of individual steps involved in the folding pathway toward the native structure; instead, they work by reducing possible unwanted aggregations of the polypeptide chain that might otherwise slow down the search for the proper intermediate and they provide a more efficient pathway for the polypeptide chain to assume the correct conformations. Chaperones are not to be confused with folding catalysts, which actually do catalyze the otherwise slow steps in the folding pathway. Examples of folding catalysts are protein disulfide isomerases and peptidyl-prolyl isomerases that may be involved in formation of disulfide bonds or interconversion between cis and trans stereoisomers, respectively. Chaperones are shown to be critical in the process of protein folding *in vivo* because they provide the protein with the aid needed to assume its proper alignments and conformations efficiently enough to become "biologically relevant". This means that the polypeptide chain could theoretically fold into its native structure without the aid of chaperones, as demonstrated by protein folding experiments conducted *in vitro*; however, this process proves to be too inefficient or too slow to exist in biological systems; therefore, chaperones are necessary for protein folding *in vivo*. Along with its role in aiding native structure formation, chaperones are shown to be involved in various roles such as protein transport, degradation, and even allow denatured proteins exposed to certain external denaturant factors an opportunity to refold into their correct native structures.

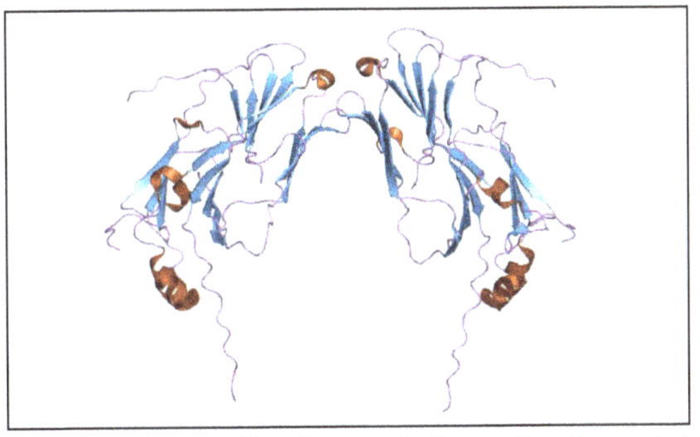

Example of a small eukaryotic heat shock protein.

A fully denatured protein lacks both tertiary and secondary structure, and exists as a so-called random coil. Under certain conditions some proteins can refold; however, in many cases, denaturation is irreversible. Cells sometimes protect their proteins against the denaturing influence of heat with enzymes known as heat shock proteins (a type of chaperone), which assist other proteins both in folding and in remaining folded. Some proteins never fold in cells at all except with the assistance of chaperones which either isolate individual proteins so that their folding is not interrupted by interactions with other proteins or help to unfold misfolded proteins, allowing them to refold into the correct native structure. This function is crucial to prevent the risk of precipitation into insoluble amorphous aggregates. The external factors involved in protein denaturation or disruption of the native state include temperature, external fields (electric, magnetic), molecular crowding, and even the limitation of space, which can have a big influence on the folding of proteins. High concentrations of solutes, extremes of pH, mechanical forces, and the presence of chemical denaturants can contribute to protein denaturation, as well. These individual factors are categorized together as stresses. Chaperones are shown to exist in increasing concentrations during times of cellular stress and help the proper folding of emerging proteins as well as denatured or misfolded ones.

Under some conditions proteins will not fold into their biochemically functional forms. Temperatures above or below the range that cells tend to live in will cause thermally unstable proteins to unfold or denature (this is why boiling makes an egg white turn opaque). Protein thermal stability is far from constant, however; for example, hyperthermophilic bacteria have been found that grow at temperatures as high as 122 °C, which of course requires that their full complement of vital proteins and protein assemblies be stable at that temperature or above.

Computational Methods for Studying Protein Folding

The study of protein folding includes three main aspects related to the prediction of protein stability, kinetics, and structure. A recent review summarizes the available computational methods for protein folding.

Energy Landscape of Protein Folding

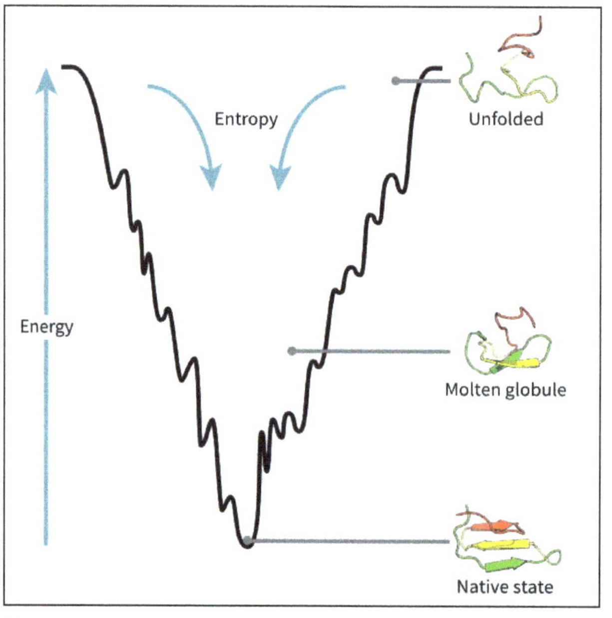

The energy funnel by which an unfolded polypeptide chain assumes its native structure.

The protein folding phenomenon was largely an experimental endeavor until the formulation of an energy landscape theory of proteins by Joseph Bryngelson and Peter Wolynes in the late 1980s and early 1990s. This approach introduced the *principle of minimal frustration*. This principle says that nature has chosen amino acid sequences so that the folded state of the protein is very stable. In addition, the undesired interactions between amino acids along the folding pathway are reduced, making the acquisition of the folded state a very fast process. Even though nature has reduced the level of *frustration* in proteins, some degree of it remains up to now as can be observed in the presence of local minima in the energy landscape of proteins. A consequence of these evolutionarily selected sequences is that proteins are generally thought to have globally "funneled energy landscapes" (coined by José Onuchic) that are largely directed toward the native state. This "folding funnel" landscape allows the protein to fold to the native state through any of a large number of pathways and intermediates, rather than being restricted to a single mechanism. The theory is supported by both computational simulations of model proteins and experimental studies, and it has been used to improve methods for protein structure prediction and design. The description of protein folding by the leveling free-energy landscape is also consistent with the 2nd law of thermodynamics. Physically, thinking of landscapes in terms of visualizable potential or total energy surfaces simply with maxima, saddle points, minima, and funnels, rather like geographic landscapes, is perhaps a little misleading. The relevant description is really a high-dimensional phase space in which manifolds might take a variety of more complicated topological forms.

The unfolded polypeptide chain begins at the top of the funnel where it may assume the largest number of unfolded variations and is in its highest energy state. Energy landscapes such as these indicate that there are a large number of initial possibilities, but only a single native state is possible; however, it does not reveal the numerous folding pathways that are possible. A different molecule of the same exact protein may be able to follow marginally different folding pathways, seeking different lower energy intermediates, as long as the same native structure is reached. Different pathways may have different frequencies of utilization depending on the thermodynamic favorability of each pathway. This means that if one pathway is found to be more thermodynamically favorable than another, it is likely to be used more frequently in the pursuit of the native structure. As the protein begins to fold and assume its various conformations, it always seeks a more thermodynamically favorable structure than before and thus continues through the energy funnel. Formation of secondary structures is a strong indication of increased stability within the protein, and only one combination of secondary structures assumed by the polypeptide backbone will have the lowest energy and therefore be present in the native state of the protein. Among the first structures to form once the polypeptide begins to fold are alpha helices and beta turns, where alpha helices can form in as little as 100 nanoseconds and beta turns in 1 microsecond.

There exists a saddle point in the energy funnel landscape where the transition state for a particular protein is found. The transition state in the energy funnel diagram is the conformation that must be assumed by every molecule of that protein if the protein wishes to finally assume the native structure. No protein may assume the native structure without first passing through the transition state. The transition state can be referred to as a variant or premature form of the native state rather than just another intermediary step. The folding of the transition state is shown to be rate-determining, and even though it exists in a higher energy state than the native fold, it greatly resembles the native structure. Within the transition state, there exists a nucleus around which the protein is able to fold, formed by a process referred to as "nucleation condensation" where the structure begins to collapse onto the nucleus.

Two models of protein folding are currently being confirmed:

- The diffusion collision model, in which first a nucleus forms, then the secondary structure, and finally these secondary structures collide and pack tightly together.
- The nucleation-condensation model, in which the secondary and tertiary structures of the protein are made at the same time.

Recent studies have shown that some proteins show characteristics of both of these folding models.

For the most part, scientists have been able to study many identical molecules folding together *en masse*. At the coarsest level, it appears that in transitioning to the native state, a given amino acid sequence takes roughly the same route and proceeds through roughly the same intermediates and transition states. Often folding involves first the establishment of regular secondary and supersecondary structures, in particular alpha helices and beta sheets, and afterward tertiary structure.

Modeling of Protein Folding

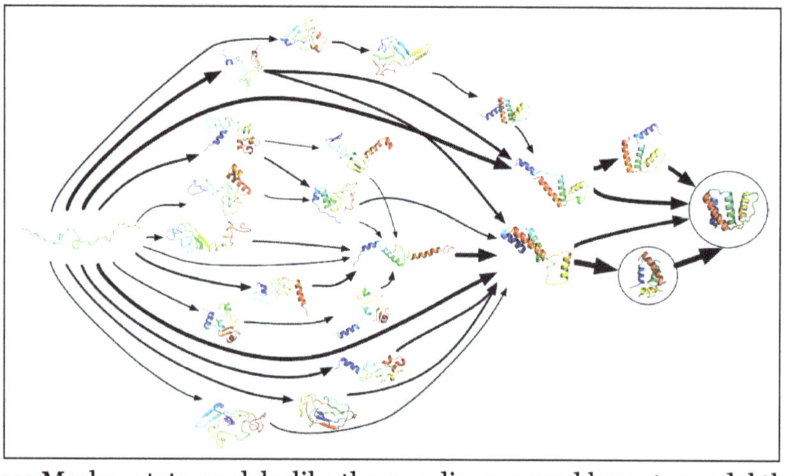

Folding@home uses Markov state models, like the one diagrammed here, to model the possible shapes and folding pathways a protein can take as it condenses from its initial randomly coiled state (left) into its native 3D structure (right).

De novo or *ab initio* techniques for computational protein structure prediction are related to, but strictly distinct from, experimental studies of protein folding. Molecular Dynamics (MD) is an important tool for studying protein folding and dynamics in silico. First equilibrium folding simulations were done using implicit solvent model and umbrella sampling. Because of computational cost, ab initio MD folding simulations with explicit water are limited to peptides and very small proteins. MD simulations of larger proteins remain restricted to dynamics of the experimental structure or its high-temperature unfolding. Long-time folding processes (beyond about 1 millisecond), like folding of small-size proteins (about 50 residues) or larger, can be accessed using coarse-grained models.

The 100-petaFLOP distributed computing project Folding@home created by Vijay Pande's group at Stanford University simulates protein folding using the idle processing time of CPUs and GPUs of personal computers from volunteers. The project aims to understand protein misfolding and accelerate drug design for disease research.

Long continuous-trajectory simulations have been performed on Anton, a massively parallel supercomputer designed and built around custom ASICs and interconnects by D. E. Shaw Research. The longest published result of a simulation performed using Anton is a 2.936 millisecond simulation of NTL9 at 355 K.

Experimental Techniques for Studying Protein Folding

While inferences about protein folding can be made through mutation studies, typically, experimental techniques for studying protein folding rely on the gradual unfolding or folding of proteins and observing conformational changes using standard non-crystallographic techniques.

X-ray Crystallography

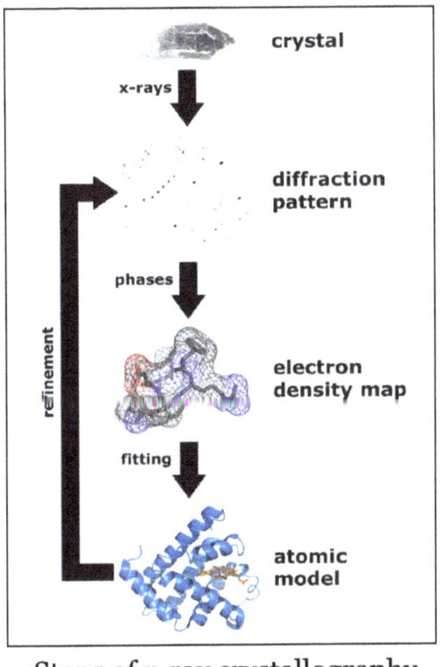

Steps of x-ray crystallography.

X-ray crystallography is one of the more efficient and important methods for attempting to decipher the three dimensional configuration of a folded protein. To be able to conduct X-ray crystallography, the protein under investigation must be located inside a crystal lattice. To place a protein inside a crystal lattice, one must have a suitable solvent for crystallization, obtain a pure protein at supersaturated levels in solution, and precipitate the crystals in solution. Once a protein is crystallized, x-ray beams can be concentrated through the crystal lattice which would diffract the beams or shoot them outwards in various directions. These exiting beams are correlated to the specific three-dimensional configuration of the protein enclosed within. The x-rays specifically interact with the electron clouds surrounding the individual atoms within the protein crystal lattice and produce a discernible diffraction pattern. Only by relating the electron density clouds with the amplitude of the x-rays can this pattern be read and lead to assumptions of the phases or phase angles involved that complicate this method. Without the relation established through a mathematical basis known as Fourier transform, the "phase problem" would render predicting the diffraction patterns very difficult. Emerging methods like multiple isomorphous replacement use the presence of a heavy metal ion to diffract the x-rays into a more predictable manner, reducing the number of variables involved and resolving the phase problem.

Fluorescence Spectroscopy

Fluorescence spectroscopy is a highly sensitive method for studying the folding state of proteins. Three amino acids, phenylalanine (Phe), tyrosine (Tyr) and tryptophan (Trp), have intrinsic fluorescence properties, but only Tyr and Trp are used experimentally because their quantum yields are high enough to give good fluorescence signals. Both Trp and Tyr are excited by a wavelength of 280 nm, whereas only Trp is excited by a wavelength of 295 nm. Because of their aromatic character, Trp and Tyr residues are often found fully or partially buried in the hydrophobic core of proteins, at the interface between two protein domains, or at the interface between subunits of oligomeric proteins. In this apolar environment, they have high quantum yields and therefore high fluorescence intensities. Upon disruption of the protein's tertiary or quaternary structure, these side chains become more exposed to the hydrophilic environment of the solvent, and their quantum yields decrease, leading to low fluorescence intensities. For Trp residues, the wavelength of their maximal fluorescence emission also depend on their environment.

Fluorescence spectroscopy can be used to characterize the equilibrium unfolding of proteins by measuring the variation in the intensity of fluorescence emission or in the wavelength of maximal emission as functions of a denaturant value. The denaturant can be a chemical molecule (urea, guanidinium hydrochloride), temperature, pH, pressure, etc. The equilibrium between the different but discrete protein states, i.e. native state, intermediate states, unfolded state, depends on the denaturant value; therefore, the global fluorescence signal of their equilibrium mixture also depends on this value. One thus obtains a profile relating the global protein signal to the denaturant value. The profile of equilibrium unfolding may enable one to detect and identify intermediates of unfolding. General equations have been developed by Hugues Bedouelle to obtain the thermodynamic parameters that characterize the unfolding equilibria for homomeric or heteromeric proteins, up to trimers and potentially tetramers, from such profiles. Fluorescence spectroscopy can be combined with fast-mixing devices such as stopped flow, to measure protein folding kinetics, generate a chevron plot and derive a Phi value analysis.

Circular Dichroism

Circular dichroism is one of the most general and basic tools to study protein folding. Circular dichroism spectroscopy measures the absorption of circularly polarized light. In proteins, structures such as alpha helices and beta sheets are chiral, and thus absorb such light. The absorption of this light acts as a marker of the degree of foldedness of the protein ensemble. This technique has been used to measure equilibrium unfolding of the protein by measuring the change in this absorption as a function of denaturant concentration or temperature. A denaturant melt measures the free energy of unfolding as well as the protein's m value, or denaturant dependence. A temperature melt measures the melting temperature (T_m) of the protein. As for fluorescence spectroscopy, circular dichroism spectroscopy can be combined with fast-mixing devices such as stopped flow to measure protein folding kinetics and to generate chevron plots.

Vibrational Circular Dichroism of Proteins

The more recent developments of vibrational circular dichroism (VCD) techniques for proteins, currently involving Fourier transform (FFT) instruments, provide powerful means for determining protein conformations in solution even for very large protein molecules. Such VCD studies of proteins are often combined with X-ray diffraction of protein crystals, FT-IR data for protein solutions in heavy water (D_2O), or *ab initio* quantum computations to provide unambiguous structural assignments that are unobtainable from CD.

Protein Nuclear Magnetic Resonance Spectroscopy

Protein folding is routinely studied using NMR spectroscopy, for example by monitoring hydrogen-deuterium exchange of backbone amide protons of proteins in their native state, which provides both the residue-specific stability and overall stability of proteins.

Dual Polarisation Interferometry

Dual polarisation interferometry is a surface-based technique for measuring the optical properties of molecular layers. When used to characterize protein folding, it measures the conformation by determining the overall size of a monolayer of the protein and its density in real time at sub-Angstrom resolution, although real-time measurement of the kinetics of protein folding are limited to processes that occur slower than ~10 Hz. Similar to circular dichroism, the stimulus for folding can be a denaturant or temperature.

Studies of Folding with High Time Resolution

The study of protein folding has been greatly advanced in recent years by the development of fast, time-resolved techniques. Experimenters rapidly trigger the folding of a sample of unfolded protein and observe the resulting dynamics. Fast techniques in use include neutron scattering, ultrafast mixing of solutions, photochemical methods, and laser temperature jump spectroscopy. Among the many scientists who have contributed to the development of these techniques are Jeremy Cook, Heinrich Roder, Harry Gray, Martin Gruebele, Brian Dyer, William Eaton, Sheena Radford, Chris Dobson, Alan Fersht, Bengt Nölting and Lars Konermann.

Proteolysis

Proteolysis is routinely used to probe the fraction unfolded under a wide range of solution conditions (e.g. Fast parallel proteolysis (FASTpp).

Optical Tweezers

Single molecule techniques such as optical tweezers and AFM have been used to understand protein folding mechanisms of isolated proteins as well as proteins with chaperones. Optical tweezers have been used to stretch single protein molecules from their C- and N-termini and unfold them to allow study of the subsequent refolding. The technique allows one to measure folding rates at single-molecule level; for example, optical tweezers have been recently applied to study folding and unfolding of proteins involved in blood coagulation. von Willebrand factor (vWF) is a protein with an essential role in blood clot formation process. It discovered – using single molecule optical tweezers measurement – that calcium-bound vWF acts as a shear force sensor in the blood. Shear force leads to unfolding of the A2 domain of vWF, whose refolding rate is dramatically enhanced in the presence of calcium. Recently, it was also shown that the simple src SH3 domain accesses multiple unfolding pathways under force.

Incorrect Protein Folding and Neurodegenerative Disease

A protein is recognized as misfolded if it cannot achieve the native state of a protein. This can be due to unwanted mutations in the amino acid sequence or could be caused through errors in the folding process. The misfolded protein typically contains beta-sheets that are organized in a polymeric arrangement known as a cross-beta structure. The abnormal proteins that are rich in beta sheet structure become partially resistant to proteolysis. Part of the reason is because beta sheets are stabilized by intermolecular interactions, therefore the misfolded proteins will have a high tendency to form oligomers and larger polymers. Also the accumulation of misfolded proteins can lead to an accumulation of protein aggregates or oligomers in the cell. The increased levels of aggregated proteins in the cell leads to formation of amyloid- like structures which can cause degenerative disorders and cell death. The amyloids are fibrillary structure that contain intermolecular hydrogen bonds, which are highly insoluble, and made from converted protein aggregates. Therefore, the proteasome pathway may not be efficient enough to degrade the misfolded proteins prior to aggregation. Misfolded proteins can interact with one another and form structured aggregates and gain toxicity through intermolecular interactions.

Aggregated proteins are associated with prion-related illnesses such as Creutzfeldt–Jakob disease, bovine spongiform encephalopathy (mad cow disease), amyloid-related illnesses such as Alzheimer's disease and familial amyloid cardiomyopathy or polyneuropathy, as well as intracytoplasmic aggregation diseases such as Huntington's and Parkinson's disease. These age onset degenerative diseases are associated with the aggregation of misfolded proteins into insoluble, extracellular aggregates and/or intracellular inclusions including cross-beta sheet amyloid fibrils. It is not completely clear whether the aggregates are the cause or merely a reflection of the loss of protein homeostasis, the balance between synthesis, folding, aggregation and protein turnover. Recently the European Medicines Agency approved the use of Tafamidis or Vyndaqel (a kinetic stabilizer of tetrameric transthyretin) for the treatment of transthyretin amyloid diseases. This suggests that the process of amyloid fibril formation (and not the fibrils themselves) causes the degeneration of post-mitotic tissue in human amyloid diseases. Misfolding and excessive degradation instead of folding and function leads to a number of proteopathy diseases such as antitrypsin-associated emphysema, cystic fibrosis and the lysosomal storage diseases, where loss of function is the origin of the disorder. While protein replacement therapy has historically been used to correct the latter disorders, an emerging approach is to use pharmaceutical chaperones to fold mutated proteins to render them functional.

Levinthal's Paradox and Kinetics

In 1969, Cyrus Levinthal noted that, because of the very large number of degrees of freedom in an unfolded polypeptide chain, the molecule has an astronomical number of possible conformations. An estimate of 3^{300} or 10^{143} was made in one of his papers. Levinthal's paradox is a thought experiment based on the observation that if a protein were folded by sequentially sampling of all possible conformations, it would take an astronomical amount of time to do so, even if the conformations were sampled at a rapid rate (on the nanosecond or picosecond scale). Based upon the observation that proteins fold much faster than this, Levinthal then proposed that a random conformational search does not occur, and the protein must, therefore, fold through a series of meta-stable intermediate states.

The duration of the folding process varies dramatically depending on the protein of interest. When studied outside the cell, the slowest folding proteins require many minutes or hours to fold primarily due to proline isomerization, and must pass through a number of intermediate states, like checkpoints, before the process is complete. On the other hand, very small single-domain proteins with lengths of up to a hundred amino acids typically fold in a single step. Time scales of milliseconds are the norm and the very fastest known protein folding reactions are complete within a few microseconds.

Protein Separation

In order to analyse different components of a mixture, their separation is necessary. A successful separation needs different separation techniques. There are several physical and chemical methods to separate the different components of a simple mixture but for a complex mixture as, mixture of amino acids or proteins, we need more advanced techniques. Chromatography is the result of advancement of separation and purification techniques. The word Chromatography comes from Greek words chroma chroma "color" and graphein "to write". Literally meaning of Chromatography is "color writing" as initially color of separated component was used for identification. This separation technique credited to a Russian botanist Dr. Michel Tswett (1903). He separated the green plant pigment and showed different components of chlorophyll in form of different colour bands on a $CaCO_3$ column through adsorption. Likewise Richard Kuhn resolves α and β isomers of carotene and proved the importance of adsorption in analytical field (The Nobel Prize in Chemistry 1938).

Column Chromatography

This is the most commonly used mode of chromatography. It can be defined as a separation process involving the uniform percolation of a liquid through a column packed with finely divided material. The stationary phase attached to a matrix (inert insoluble support) is packed in a glass or metal column and mobile phase is passed through it either by gravitational flow or with help of a pump. The selected stationary phase retards the movement of certain components of mobile phase which leads to effective separation of different components. This retardation in movement may be either due to direct interaction of solute component with stationary phase or indirect adsorption of solute component on the surface of stationary phase.

A typical column chromatographic system uses liquid mobile phase, consist of a column, a mobile phase reservoir and delivery system (pump), a detector for identification of separated analytes as they emerge in the effluent from the column, a recorder to maintain the record of analytes in combination with detector and a fraction collector to collect each analyte separately.

The pictorial representation of detector response as a function of elution time or volume is known as chromatogram which consists of a series of peaks, representing the individual analytes. After application on the column the time require to elute by an analyte is known as retention time (t_R) of that analyte. The retention time of any analyte has two components. The first one is time taken by that analyte to cross the column through free spaces between matrixes. This volume is referred as void volume (V_o) and the time taken is called dead time (t_M). The value of dead time (t_M) will be same for all analytes and can be measured by the analyte which does not show any interaction with stationary phase. The second component is the time the analyte retained by the stationary phase and known as adjusted retention time ($t_{R'}$). This is the characteristic of analyte and can be given by;

$$t_{R'} = t_R - t_M$$

The additional time taken by an analyte to elute from the column relative to an excluded analyte that does not interact with the stationary phase is known as capacity factor (k').

Thus, Capacity factor $(k') = t_R - t_M / t_M = t_R' / t_M$ (capacity factor has no units).

Column efficiency and resolution: The efficiency of a chromatographic column is a measure of the capacity of the column to restrain peak dispersion and thus, provide high resolution. The higher the efficiency, the more the peak dispersion is restrained, and the better the column. The column efficiency will vary with the retention of the peak. In capillary columns, the efficiency generally falls as the retention increases and for a packed column the efficiency generally increases with retention.

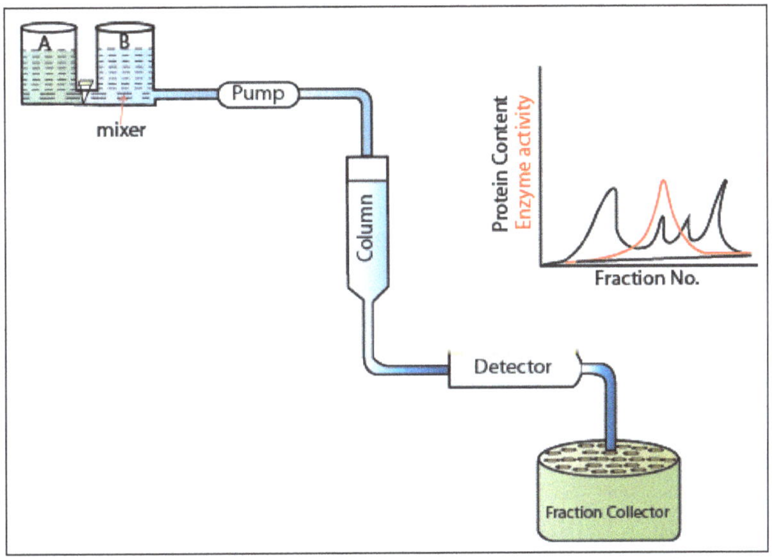

A pictorial representation of column chromatography.

The expression for calculating the column efficiency (N) can be derived from the plate theory. Column efficiency is measured in theoretical plates (from the plate theory) and is taken as 16 times the square of the ratio of the retention time (the time between the injection point and the peak maximum) to the peak width at the points of inflection.

$$N = 16 \, t_{R^2} / w^2$$

The height equivalent to the theoretical plate (HETP) or the variance per unit Length of a column is calculated as the ratio of the column length to the column efficiency (number of plates). If the length of the column is L , then the HETP (height equivalent to the theoretical plate) is;

$$HETP = L / N$$

Note: The plate model supposes that the chromatographic column contains a large number of separate layers, called theoretical plates . Separate equilibrations of the sample between the stationary and mobile phase occur in these "plates". The analyte moves down the column by transfer of equilibrated mobile phase from one plate to the next. It is important to remember that the plates do not really exist ; they are a figment of the imagination that helps us understand the processes at work in the column. They also serve as a way of measuring column efficiency, either by stating the number of theoretical plates in a column, N (the more plates the better), or by stating the plate height; (the height equivalent to a theoretical plate) (Smaller the HEPT, better the column efficiency).

The number of theoretical plates that a real column possesses can be found by examining a chromatographic peak after elution;

$$N = \mathbf{5.55\, t_{R^2} / w_{1/2^2}} \text{ (where } w_{1/2} \text{ is the peak width at half – height)}$$

Resolution of chromatographic column is the ability to resolve one analyte peak to other. Resolution can be defined as the ratio of the difference in retention time between the two peaks to the mean of their base widths (w_{av}).

$$\mathbf{R_S = 2(t_{RA} - t_{RB}) / w_A + w_B = dt_R / w_{av}}$$

On the basis of the pressure generated inside the column the liquid column chromatography can be further subdivided as:

- Low pressure liquid chromatography (LPLC) - < 5 Bar.
- Medium pressure liquid chromatography (MPLC) – 6 to 50 Bar.
- High pressure liquid chromatography (HPLC) - > 50 Bar.

In low pressure chromatography, stationary phase are mainly polysaccharides which are mechanically weak. Therefore, even if such particles are produced with small diameters, they would not be sufficiently strong to withstand the high pressure required for high resolution chromatography. Thus, in HPLC silica based particles (5-10 μm) are used. Smaller size of particle (large surface area) reduces size of theoretical chromatographic plat (thus number of plate in a given length is more). Higher number of chromatographic plate results in better resolution.

Types of column chromatography: On the basis of type of interaction between the stationary and the mobile phases the column chromatography can be divided into following types:

1. Adsorption chromatography.
2. Partition chromatography.
3. Ion-Exchange chromatography.
4. Molecular Exclusion (Gel Filtration) chromatography.
5. Affinity chromatography.

Adsorption Chromatography

Adsorption is a surface phenomenon in which molecules get attached to column particle. The molecule is called adsorbate while column particle is adsorbent. Molecules can be separated based on their adsorption properties. Many inert materials can be used as adsorbents. e.g. alumina, charcoal, calcium carbonate.

Partition Chromatography

Chromatography is a non destructive separation technique for successful separation and purification of individual components of a complex mixture which can not be separated by usual methods due to their similar physical and chemical properties. The basic principal of any chromatography is, how a compound get distributed in between two immiscible phases, a stationary phase and a mobile phase. Those components have higher affinity for the stationary phase are retained longer in the system than those that are distributed selectively in the mobile phase. As a consequence, solutes are eluted from the system as local concentrations in the mobile phase in the order of their increasing distribution coefficients with respect to the stationary phase which leads the separation of different components of mixture.

An analyte remains in equilibrium between the two phases,

$$A_{mobile} \rightleftarrows A_{stationary}$$

The equilibrium constant, K_d is termed the partition coefficient or distribution coefficient, defined as the molar concentration of analyte in the stationary phase divided by the molar concentration of the analyte in the mobile phase. It explains about the distribution of a compound in between two phases. Suppose there are two phases A & B (immiscible to each other) and these are present together. When a compound X is mixed, it distributes itself in these two phases and the concentration of compound in phase A is X_A and in phase B is X_B. The distribution coefficient of X can be expressed as,

$$K_d = X_A / X_B \text{ } (K_d \text{ is a constant at a particular temperature})$$

Factors Affecting the Magnitude of the Distribution Coefficient (K_d)

The magnitude of (K_d) is determined by the relative affinity of the solute for the two phases. Those solutes interacting more strongly with the stationary phase will exhibit a high distribution coefficient and will be retained longer in the chromatographic system. Molecular interaction results from intermolecular forces of which there are three basic types. (a) Dispersion forces (b) Polar forces (c) Ionic forces.

If the compound X has same affinity for both the phase, it gets distributed equally in both the phases but compound Y has differential affinity for two phases and accordingly gets distributed in stationary and mobile phase.

Protein Chemistry

The central dogma of life places translation process and protein products at the final stage. Protein play most important role in determining the role of the cell in the system. The DNA sequence determines the sequence of amino acids in the protein, but it is actually the modifications in the proteins resulting from alternate splicing and post posttranslational modification, which finally dictates the physiological function of the protein. The structure and the function of the proteins are closely associated. In fact, it is the 3D structure of the protein, which governs the function of the protein. Hence an approach should be available so that all information regarding proteins can be obtained in high throughput manner.

Proteomics differs from the conventional protein chemistry approach of identification, structural and functional elucidation of proteins, in many aspects. Proteomics aims to decipher the protein properties from a global perspective, not taking one protein at a time. This approach becomes a holistic model, while studying the system, because in a system the protein in question is not alone. It is associated and interacting with several other proteins and bio-molecules, which together determine its role in the system. These two approaches, protein chemistry and proteomics, also differ in the techniques that are employed for these approaches as well as scale of usability. While traditional protein chemistry also employed techniques like electrophoresis and mass spectrometry, it is the advancement of these techniques, which led to the divergence from protein chemistry to proteomics. The sensitivity, resolvability, robustness and the high throughput approaches made the transition possible. The completion of genome sequence projects and advancement in bioinformatics and microarrays were also major catalysts for the advancement of proteomics. These advancements in area of protein chemistry and biology gave rise to this new discipline, proteomics, and with continuous evolution of several advanced technologies, this field is continuously advancing.

Evolution of Proteomics

The number of protein coding genes in the human genome is approximately 5% of the entire genome. It is an astonishing fact that how these 5% of the genes account for the entire diversity of proteins in the cell. It is the dynamic properties of proteins, which lead to several diseases. For example, cell cycle regulating proteins or cyclin dependent kinases phosphorylate activating the cyclins, which promote the cell cycle by allowing DNA replication to pass through the check points. Proteomics emerged from protein chemistry solely because of the advancement in the existing technologies.

Advancement in Mass Spectrometry

Mass spectrometry has been into existence during last several decades, since the days of protein chemistry. Mass spectrometry ideally measures the mass of an analyte by producing charged molecular species in vacuum, and their separation by magnetic and electric fields based on mass to charge (m/z) ratio. The pre-requisite for a mass spectrometry analysis is the generation of ionized particles. However, proteins being large soluble polymers of amino acids could not be ionized by the conventional gas chromatography without fragmenting it into constituent amino acids. This for the time being limited the usage of mass spectrometry in protein chemistry. However, with major discoveries of soft ionization techniques like MALDI and ESI, the mass spectrometry became a robust analytical technique for protein study.

With advancement in ionization techniques, sophisticated mass analyzers and detectors, the mass analysis of as low as 1 ppm with excellent resolving power is possible. The sensitivity of the mass spectrometry also increased, making it possible to detect the attomole of substances. The tandem mass spectrometry involving two mass analyzers like Tof-Tof/ Q-Tof etc. helped in protein sequencing. The Edman degradation process of amino acid sequencing though was extremely useful, was not high throughput as it could sequence a stretch of maximum 40 amino acids, whereas mass spectrometry could do it for large proteins. The ability of mass spectrometry for relative and absolute quantitation of proteins by employing iTRAQ, ICAT and SILAC labels have definitely advanced the field of quantitative proteomics.

Advancement in Electrophoresis

Electrophoresis refers to the process of separation of charged particles under the influence of an external electric field. Electrophoretic separation of proteins had been used widely for quite a long time; however, recent advances in electrophoretic techniques in the form of protein separation, staining and detection have advanced this technique for further usage in proteomics.

The first advancement came in the transition from tube gels to immobilized pH gradient strips. Earlier, isoelectric focusing was performed using tube gels, which had biggest disadvantage of their stability. The pH would often change and result in erroneous results. The tube gels suffered from extremes of variations, discontinuity in the entire gel, which led to its breakage when concentrated samples were added. Prof. Angelika Gorg has made a remarkable contribution in the field of electrophoresis by substituting the tube gels with immobilized pH gradient strips. These were Acrylamide coated plastic strips containing immobilins of various pH spread across them. The biggest advantage of these strips was in the stability of the pH ampholytes inside the gel. As a result gel to gel variations got reduced and the physical stability of the strips were enhanced, as much as, they can be stored after the first dimension for a week at -200C before the second dimension can be done.

Staining techniques also increased the usability of gel electrophoresis in proteomic studies. The conventional coomassie brilliant blue dye had the limitation of 40 mg protein requirement. Although silver staining increased the sensitivity but it had its own limitation in terms of background staining and incompatibility with mass spectrometry for the protein identification. To overcome these problems, cyanine dyes were introduced for staining which have extreme sensitivity as well as the specificity. The cyanine dyes interact with the protein with their hydroxysuccinamide residues and fluoresce when subjected to light of appropriate wavelength. This property of cyanine dyes is utilized into developing a technique known as DIGE (Difference in gel electrophoresis). The biggest advantage of DIGE was in its ability to resolve gel-to-gel variation. Since an internal standard is included in the gel, there is no need to perform electrophoresis on many samples. The sensitivity of the dyes, coupled with the latest software could thus even check the smallest amount of differential expression level of a protein with reliability. Although electrophoretic techniques such as 2DE is now considered to be primitive in proteomics; however, quantitative approaches such as 2D-DIGE still hold good place for quantitative proteomic analysis.

Protein Mass Spectrometry

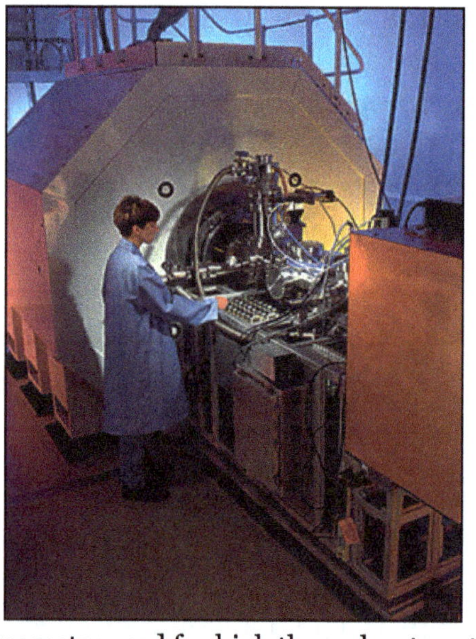

A mass spectrometer used for high throughput protein analysis.

Protein mass spectrometry refers to the application of mass spectrometry to the study of proteins. Mass spectrometry is an important method for the accurate mass determination and characterization of proteins, and a variety of methods and instrumentations have been developed for its many uses. Its applications include the identification of proteins and their post-translational modifications, the elucidation of protein complexes, their subunits and functional interactions, as well as the global measurement of proteins in proteomics. It can also be used to localize proteins to the various organelles, and determine the interactions between different proteins as well as with membrane lipids.

The two primary methods used for the ionization of protein in mass spectrometry are electrospray ionization (ESI) and matrix-assisted laser desorption/ionization (MALDI). These ionization techniques are used in conjunction with mass analyzers such as tandem mass spectrometry. In general, the protein are analyzed either in a "top down" approach in which proteins are analyzed intact, or a "bottom-up" approach in which protein are first digested into fragments. An intermediate "middle-down" approach in which larger peptide fragments are analyzed may also sometimes be used.

History

The application of mass spectrometry to study proteins became popularized in the 1980s after the development of MALDI and ESI. These ionization techniques have played a significant role in the characterization of proteins. (MALDI) Matrix-assisted laser desorption ionization was coined in the late 80's by Franz Hillenkamp and Michael Karas. Hillenkamp, Karas and their fellow researchers were able to ionize the amino acid alanine by mixing it with the amino acid tryptophan and irradiated with a pulse 266 nm laser. Though important, the breakthrough did not come until 1987. In 1987, Koichi Tanaka used the "ultra fine metal plus liquid matrix method" and ionized biomolecules the size of 34,472 Da protein carboxypeptidase-A.

In 1968, Malcolm Dole reported the first use of electrospray ionization with mass spectrometry. Around the same time MALDI became popularized, John Bennett Fenn was cited for the development of electrospray ionization. Koichi Tanaka received the 2002 Nobel Prize in Chemistry alongside John Fenn, and Kurt Wüthrich "for the development of methods for identification and structure analyses of biological macromolecules." These ionization methods have greatly facilitated the study of proteins by mass spectrometry. Consequently, protein mass spectrometry now plays a leading role in protein characterization.

Methods and Approaches

Techniques

Mass spectrometry of proteins requires that the proteins in solution or solid state be turned into an ionized form in the gas phase before they are injected and accelerated in an electric or magnetic field for analysis. The two primary methods for ionization of proteins are electrospray ionization (ESI) and matrix-assisted laser desorption/ionization (MALDI). In electrospray, the ions are created from proteins in solution, and it allows fragile molecules to be ionized intact, sometimes preserving non-covalent interactions. In MALDI, the proteins are embedded within a matrix normally in a solid form, and ions are created by pulses of laser light. Electrospray produces more multiply-charged ions than MALDI, allowing for measurement of high mass protein and better fragmentation for identification, while MALDI is fast and less likely to be affected by contaminants, buffers and additives.

Whole-protein mass analysis is primarily conducted using either time-of-flight (TOF) MS, or Fourier transform ion cyclotron resonance (FT-ICR). These two types of instrument are preferable here because of their wide mass range, and in the case of FT-ICR, its high mass accuracy. Electrospray ionization of a protein often results in generation of multiple charged species of 800 < m/z < 2000 and the resultant spectrum can be deconvoluted to determine the protein's average mass to within 50 ppm or better using TOF or ion-trap instruments.

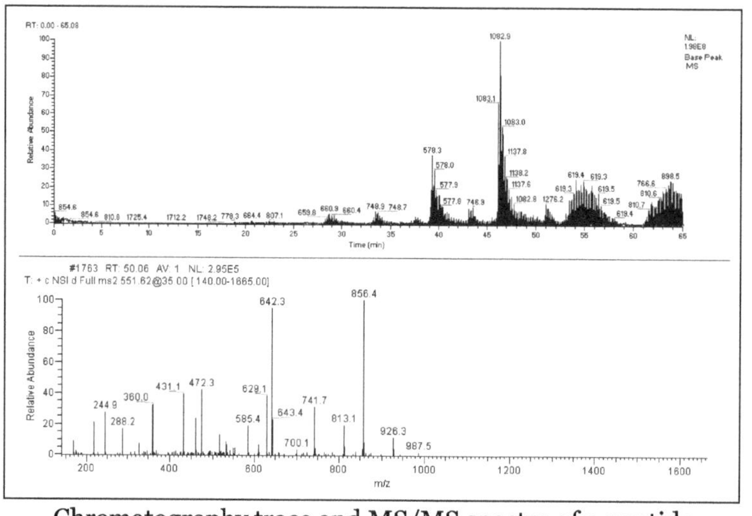

Chromatography trace and MS/MS spectra of a peptide.

Mass analysis of proteolytic peptides is a popular method of protein characterization, as cheaper instrument designs can be used for characterization. Additionally, sample preparation is easier once whole proteins have been digested into smaller peptide fragments. The most widely used instrument for peptide mass analysis are the MALDI-TOF instruments as they permit the acquisition of peptide mass fingerprints (PMFs) at high pace (1 PMF can be analyzed in approx. 10 sec). Multiple stage quadrupole-time-of-flight and the quadrupole ion trap also find use in this application.

Tandem mass spectrometry (MS/MS) is used to measure fragmentation spectra and identify proteins at high speed and accuracy. Collision-induced dissociation is used in mainstream applications to generate a set of fragments from a specific peptide ion. The fragmentation process primarily gives rise to cleavage products that break along peptide bonds. Because of this simplicity in fragmentation, it is possible to use the observed fragment masses to match with a database of predicted masses for one of many given peptide sequences. Tandem MS of whole protein ions has been investigated recently using electron capture dissociation and has demonstrated extensive sequence information in principle but is not in common practice.

Approaches

In keeping with the performance and mass range of available mass spectrometers, two approaches are used for characterizing proteins. In the first, intact proteins are ionized by either of the two techniques described above, and then introduced to a mass analyzer. This approach is referred to as "top-down" strategy of protein analysis as it involves starting with the whole mass and then pulling it apart. The top-down approach however is mostly limited to low throughput single-protein studies due to issues involved in handling whole proteins, their heterogeneity and the complexity of their analyses.

In the second approach, referred to as the "bottom-up" MS, proteins are enzymatically digested into smaller peptides using a protease such as trypsin. Subsequently, these peptides are introduced into the mass spectrometer and identified by peptide mass fingerprinting or tandem mass spectrometry. Hence, this approach uses identification at the peptide level to infer the existence of proteins pieced back together with *de novo* repeat detection. The smaller and more uniform fragments are easier to analyze than intact proteins and can be also determined with high accuracy, this "bottom-up" approach is therefore the preferred method of studies in proteomics. A further approach that is beginning to be useful is the intermediate "middle-down" approach in which proteolytic peptides larger than the typical tryptic peptides are analyzed.

Protein and Peptide Fractionation

Proteins of interest are usually part of a complex mixture of multiple proteins and molecules, which co-exist in the biological medium. This presents two significant problems. First, the two ionization techniques used for large molecules only work well when the mixture contains roughly equal amounts of constituents, while in biological samples, different proteins tend to be present in widely differing amounts. If such a mixture is ionized using electrospray or MALDI, the more abundant species have a tendency to "drown" or suppress signals from less abundant ones. Second, mass spectrum from a complex mixture is very difficult to interpret due to the overwhelming number of mixture components. This is exacerbated by the fact that enzymatic digestion of a protein gives rise to a large number of peptide products.

In light of these problems, the methods of one- and two-dimensional gel electrophoresis and high performance liquid chromatography are widely used for separation of proteins. The first method fractionates whole proteins via two-dimensional gel electrophoresis. The first-dimension of 2D gel is isoelectric focusing (IEF). In this dimension, the protein is separated by its isoelectric point (pI) and the second-dimension is SDS-polyacrylamide gel electrophoresis (SDS-PAGE). This dimension separates the protein according to its molecular weight. Once this step is completed in-gel digestion occurs. In some situations, it may be necessary to combine both of these techniques. Gel spots identified on a 2D Gel are usually attributable to one protein. If the identity of the protein is desired, usually the method of in-gel digestion is applied, where the protein spot of interest is excised, and digested proteolytically. The peptide masses resulting from the digestion can be determined by mass spectrometry using peptide mass fingerprinting. If this information does not allow unequivocal identification of the protein, its peptides can be subject to tandem mass spectrometry for *de novo* sequencing.

Small changes in mass and charge can be detected with 2D-PAGE. The disadvantages with this technique are its small dynamic range compared to other methods, some proteins are still difficult to separate due to their acidity, basicity, hydrophobicity, and size (too large or too small).

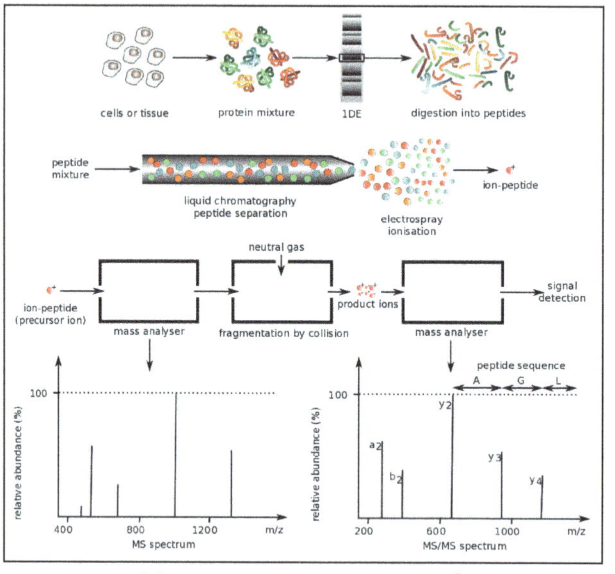

Mass spectrometry protocol.

The second method, high performance liquid chromatography is used to fractionate peptides after enzymatic digestion. Characterization of protein mixtures using HPLC/MS is also called shotgun proteomics and MuDPIT (Multi-Dimensional Protein Identification Technology). A peptide mixture that results from digestion of a protein mixture is fractionated by one or two steps of liquid chromatography. The eluent from the chromatography stage can be either directly introduced to the mass spectrometer through electrospray ionization, or laid down on a series of small spots for later mass analysis using MALDI.

Applications

Protein Identification

There are two main ways MS is used to identify proteins. Peptide mass fingerprinting uses the masses of proteolytic peptides as input to a search of a database of predicted masses that would arise from digestion of a list of known proteins. If a protein sequence in the reference list gives rise to a significant number of predicted masses that match the experimental values, there is some evidence that this protein was present in the original sample. Purification steps therefore limit the throughput of the peptide mass fingerprinting approach. Peptide mass fingerprinting can be achieved with MS/MS.

MS is also the preferred method for the identification of post-translational modifications in proteins as it is more advantageous than other approaches such as the antibody-based methods.

De Novo (Peptide) Sequencing

De novo peptide sequencing for mass spectrometry is typically performed without prior knowledge of the amino acid sequence. It is the process of assigning amino acids from peptide fragment masses of a protein. *De novo* sequencing has proven successful for confirming and cxpanding upon results from database searches.

As *de novo* sequencing is based on mass and some amino acids have identical masses (e.g. leucine and isoleucine), accurate manual sequencing can be difficult. Therefore, it may be necessary to utilize a sequence homology search application to work in tandem between a database search and *de novo* sequencing to address this inherent limitation.

Database searching has the advantage of quickly identifying sequences, provided they have already been documented in a database. Other inherent limitations of database searching include sequence modifications/mutations (some database searches do not adequately account for alterations to the 'documented' sequence, thus can miss valuable information), the unknown (if a sequence is not documented, it will not be found), false positives, and incomplete and corrupted data.

An annotated peptide spectral library can also be used as a reference for protein/peptide identification. It offers the unique strength of reduced search space and increased specificity. The limitations include spectra not included in the library will not be identified, spectra collected from different types of mass spectrometers can have quite distinct features, and reference spectra in the library may contain noise peaks, which may lead to false positive identifications. A number of different algorithmic approaches have been described to identify peptides and proteins from tandem mass spectrometry (MS/MS), peptide *de novo* sequencing and sequence tag-based searching.

Protein Quantitation

Several recent methods allow for the quantitation of proteins by mass spectrometry (quantitative proteomics). Typically, stable (e.g. non-radioactive) heavier isotopes of carbon (^{13}C) or nitrogen (^{15}N) are incorporated into one sample while the other one is labeled with corresponding light isotopes (e.g. ^{12}C and ^{14}N). The two samples are mixed before the analysis. Peptides derived from the different samples can be distinguished due to their mass difference. The ratio of their peak intensities corresponds to the relative abundance ratio of the peptides (and proteins). The most popular methods for isotope labeling are SILAC (stable isotope labeling by amino acids in cell culture), trypsin-catalyzed ^{18}O labeling, ICAT (isotope coded affinity tagging), iTRAQ (isobaric tags for relative and absolute quantitation). "Semi-quantitative" mass spectrometry can be performed without labeling of samples. Typically, this is done with MALDI analysis (in linear mode). The peak intensity, or the peak area, from individual molecules (typically proteins) is here correlated to the amount of protein in the sample. However, the individual signal depends on the primary structure of the protein, on the complexity of the sample, and on the settings of the instrument. Other types of "label-free" quantitative mass spectrometry, uses the spectral counts (or peptide counts) of digested proteins as a means for determining relative protein amounts.

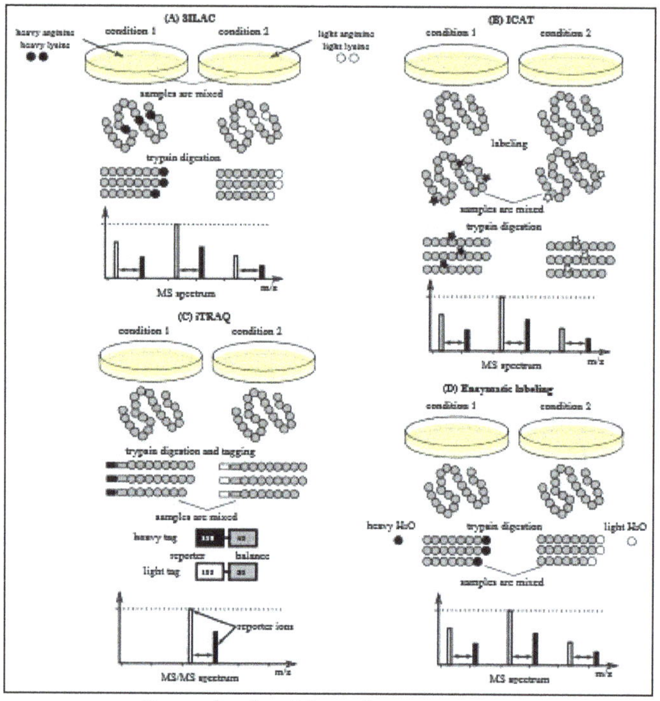

Quantitative Mass Spectrometry.

Protein Structure Determination

Characteristics indicative of the 3-dimensional structure of proteins can be probed with mass spectrometry in various ways. By using chemical crosslinking to couple parts of the protein that are close in space, but far apart in sequence, information about the overall structure can be inferred. By following the exchange of amide protons with deuterium from the solvent, it is possible to probe the solvent accessibility of various parts of the protein. Hydrogen-deuterium exchange mass spectrometry has been used to study proteins and their conformations for over 20 years. This type of protein structural analysis can be suitable for proteins that are challenging for other structural methods. Another interesting avenue in protein structural studies is laser-induced covalent labeling. In this technique, solvent-exposed sites of the protein are modified by hydroxyl radicals. Its combination with rapid mixing has been used in protein folding studies.

Biomarkers

The FDA defines a biomarker as, "A characteristic that is objectively measured and evaluated as an indicator of normal biologic processes, pathogenic processes, or pharmacologic responses to a therapeutic intervention". It is hypothesized that mass spectrometry enables the discovery of candidates for biomarkers.

Proteogenomics

In what is now commonly referred to as proteogenomics, peptides identified with mass spectrometry are used for improving gene annotations (for example, gene start sites) and protein annotations. Parallel analysis of the genome and the proteome facilitates discovery of post-translational modifications and proteolytic events, especially when comparing multiple species.

Quantitative Proteomics

Quantitative proteomics is an analytical chemistry technique for determining the amount of proteins in a sample. Rather than just providing lists of proteins identified in a certain sample, quantitative proteomics yields information about differences between samples. For example, this approach can be used to compare samples from healthy and diseased patients. The methods for protein identification are identical to those used in general (i.e. qualitative) proteomics, but include quantification as an additional dimension. Quantitative proteomics is mainly performed by two-dimensional gel electrophoresis (2-DE) or mass spectrometry (MS). However, a recent developed method of Quantitative Dot Blot (QDB) analysis is able to measure both the absolute and relative quantity of an individual proteins in the sample in high throughput format, thus open a new direction for proteomic research. In contrast to 2-DE, which requires MS for the downstream protein identification, MS technology can identify and quantify the changes.

Technologies

Mass spectrometry (MS) and two-dimensional gel electrophoresis (2-DE) represent the main technologies for quantitative proteomics with advantages and disadvantages. 2-DE provides information about the protein quantity, charge, and mass of the intact protein. It has limitations for the analysis of proteins larger than 150 kDa or smaller than 5kDa and low solubility proteins. Quantitative MS has higher sensitivity but does not provide information about the intact protein.

Classical 2-DE based on post-electrophoretic dye staining has limitations: at least three technical replicates are required to verify the reproducibility. Difference gel electrophoresis (DIGE) uses fluorescence-based labeling of the proteins prior to separation has increased the precision of quantification as well as the sensitivity in the protein detection. Therefore, DIGE represents the current main approach for the 2-DE based study of proteomes.

For quantitative MS, a commonly applied approach is isotope-coded affinity tags (ICAT), which uses two reagents with heavy and light isotopes, respectively, and a biotin affinity tag to modify cysteine containing peptides. This technology has been used to label whole *Saccharomyces cerevisiae* cells, and, in conjunction with mass spectrometry, helped lay the foundation of quantitative proteomics.

Discovery and Targeted Proteomics

Strategies to improve the sensitivity and scope of proteomic analysis often require large sample quantities and multi-dimensional fractionation, which sacrifices throughput. Alternatively, efforts to improve the sensitivity and throughput of protein quantification limit the number of peptides that can be monitored per MS run. For this reason, proteomics research is typically divided into two categories: discovery and targeted proteomics. Discovery proteomics optimizes protein identification by spending more time and effort per sample and reducing the number of samples analyzed. In contrast, targeted proteomics strategies limit the number of features that will be monitored and then optimize the chromatography, instrument tuning and acquisition methods to achieve the highest sensitivity and throughput for hundreds or thousands of samples.

Relative and Absolute Quantification

Mass spectrometry is not inherently quantitative because of differences in the ionization efficiency and/or detectability of the many peptides in a given sample, which has sparked the development of methods to determine relative and absolute abundance of proteins in samples. The intensity of a peak in a mass spectrum is not a good indicator of the amount of the analyte in the sample, although differences in peak intensity of the *same* analyte between multiple samples accurately reflect relative differences in its abundance.

Label-free Quantification

One approach for relative quantification is to separately analyze samples by MS and compare the spectra to determine peptide abundance in one sample relative to another, as in Label-free quantification strategies.

Stable Isotope Labels

An approach for relative quantification that is more costly and time-consuming, though less sensitive to experimental bias than label-free quantification, entails labeling the samples with stable isotope labels that allow the mass spectrometer to distinguish between identical proteins in separate samples. One type of label, isotopic tags, consist of stable isotopes incorporated into protein crosslinkers that causes a known mass shift of the labeled protein or peptide in the mass spectrum. Differentially labeled samples are combined and analyzed together, and the differences in the peak intensities of the isotope pairs accurately reflect difference in the abundance of the corresponding proteins.

Absolute proteomic quantification using isotopic peptides entails spiking known concentrations of synthetic, heavy isotopologues of target peptides into an experimental sample and then performing LC-MS/MS. As with relative quantification using isotopic labels, peptides of equal chemistry co-elute and are analyzed by MS simultaneously.

Unlike relative quantification, though, the abundance of the target peptide in the experimental sample is compared to that of the heavy peptide and back-calculated to the initial concentration of the standard using a pre-determined standard curve to yield the absolute quantification of the target peptide.

Relative quantification methods include isotope-coded affinity tags (ICAT), isobaric labeling (tandem mass tags (TMT) and isobaric tags for relative and absolute quantification (iTRAQ)), label-free quantification Metal-coded tags (MeCAT), N-terminal labelling, stable isotope labeling with amino acids in cell culture (SILAC), and Terminal amine isotopic labeling of substrates (TAILS).

Absolute quantification is performed using selected reaction monitoring (SRM).

MeCAT can be used in combination with element mass spectrometry ICP-MS allowing first-time absolute quantification of the metal bound by MeCAT reagent to a protein or biomolecule. Thus it is possible to determine the absolute amount of protein down to attomol range using external calibration by metal standard solution. It is compatible to protein separation by 2D electrophoresis and chromatography in multiplex experiments. Protein identification and relative quantification can be performed by MALDI-MS/MS and ESI-MS/MS.

Mass spectrometers have a limited capacity to detect low-abundance peptides in samples with a high dynamic range. The limited duty cycle of mass spectrometers also restricts the collision rate, resulting in an undersampling Sample preparation protocols represent sources of experimental bias.

References

- Gromer S, Urig S, Becker K (January 2004). "The thioredoxin system--from science to clinic". Medicinal Research Reviews. 24 (1): 40–89. PMID 14595672. doi:10.1002/med.10051
- Xie J, Schultz PG (December 2005). "Adding amino acids to the genetic repertoire". Current Opinion in Chemical Biology. 9 (6): 548–54. PMID 16260173. doi:10.1016/j.cbpa.2005.10.011
- Hausman, Robert E., Cooper, Geoffrey M. (2004). The cell: a molecular approach. Washington, D.C: ASM Press. p. 51. ISBN 0-87893-214-3
- Pratt C, Cornely K (2004). "Thermodynamics". Essential Biochemistry. Wiley. ISBN 978-0-471-39387-0. Retrieved 03, March 2020
- Alberts B, Johnson A, Lewis J, Raff M, Roberts K, Walters P (2002). "The Shape and Structure of Proteins". Molecular Biology of the Cell; Fourth Edition. New York and London: Garland Science. ISBN 0-8153-3218-1
- "Structural Biochemistry/Proteins/Protein Folding - Wikibooks, open books for an open world". en.wikibooks.org. Retrieved 11, May 2020
- Hatem, Salama Mohamed Ali (2006). "Gas chromatographic determination of Amino Acid Enantiomers in tobacco and bottled wines". University of Giessen. Retrieved 17, July 2020

- Cowtan K (2001). “Phase Problem in X-ray Crystallography, and Its Solution” (PDF). Encyclopedia of Life Sciences. Macmillan Publishers Ltd, Nature Publishing Group. Retrieved 01, January 2020
- Phillips SM (May 2014). “A brief review of critical processes in exercise-induced muscular hypertrophy”. Sports Medicine. 44 Suppl 1: S71–7. PMC 4008813 . PMID 24791918. doi:10.1007/s40279-014-0152-3

Chapter 4

Primary Proteins

Cyanovirin-N is a protein that displays virucidal activity against several viruses. Blood proteins are present in blood plasma and are also known as plasma proteins. Globular proteins are water-soluble, spherical proteins which are fairly common, hemoglobin being one of them. This chapter delves into the various proteins studied in proteomics and covers the aforementioned proteins in a comprehensive manner.

Cyanovirin-N

d Cyanovirin-N (CV-N) is a protein produced by the cyanobacterium *Nostoc ellipsosporum* that displays virucidal activity against several viruses, including human immunodeficiency virus (HIV). The virucidal activity of CV-N is mediated through specific high-affinity interactions with the viral surface envelope glycoproteins gp120 and gp41, as well as to high-mannose oligosaccharides found on the HIV envelope. In addition, CV-N is active against rhinoviruses, human parainfluenza virus, respiratory syncytial virus, and enteric viruses. The virucidal activity of CV-N against influenza virus is directed towards viral haemagglutinin. CV-N has a complex fold composed of a duplication of a tandem repeat of two homologous motifs comprising three-stranded beta-sheet and beta-hairpins.

Professor Julian Ma of St George's Hospital, South London, has a project in Kent, England to use genetically modified tobacco plants to produce the Cyanovirin and from this produce a cream which could be used to prevent HIV infection.

Cyanovirin a protein with a highly complicated structure, it binds to sugars attached to HIV,it envelopes the protein and prevents it from binding to the mucosal cell surfaces in the Vagina and Rectum, this compound is also active against herpes viruses.

"development of cyanovirin has been exceedingly slow-paced. The chief of the NCI cyanovirin program, Michael Boyd, described it as "languishing." Apparently the NCI's production facilities, based on genetically manipulated cell cultures, have been diverted to other projects that the agency considers of higher priority. This is unfortunate: cyanovirin is of particular interest because of its relative safety. It is 10,000 times more toxic to HIV than it is to cells."

This protein may use the morpheein model of allosteric regulation.

Intrinsically Disordered Proteins

An intrinsically disordered protein (IDP) is a protein that lacks a fixed or ordered three-dimensional structure. IDPs cover a spectrum of states from fully unstructured to partially structured and include random coils, (pre-)molten globules, and large multi-domain proteins connected by flexible linkers. They constitute one of the main types of protein (alongside globular, fibrous and membrane proteins).

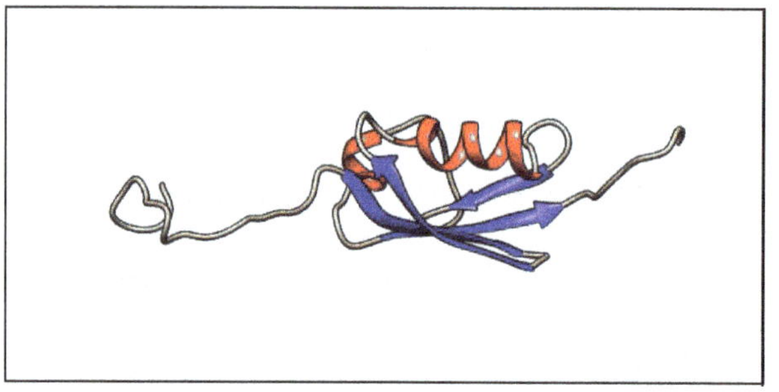

Conformational flexibility in SUMO-1 protein (PDB:1a5r). The central part shows relatively ordered structure. Conversely, the N- and C-terminal regions (left and right, respectively) show 'intrinsic disorder', although a short helical region persists in the N-terminal tail. Ten alternative NMR models were morphed. Secondary structure elements: α-helices (red), β-strands (blue arrows).

The discovery of IDPs has challenged the traditional protein structure paradigm, that protein function depends on a fixed three-dimensional structure. This dogma has been challenged over the last decades by increasing evidence from various branches of structural biology, suggesting that protein dynamics may be highly relevant for such systems. Despite their lack of stable structure, IDPs are a very large and functionally important class of proteins. In some cases, IDPs can adopt a fixed three-dimensional structure after binding to other macromolecules. Overall, IDPs are different from structured proteins in many ways and tend to have distinct properties in terms of function, structure, sequence, interactions, evolution and regulation.

History

An ensemble of NMR structures of the Thylakoid soluble phosphoprotein TSP9, which shows a largely flexible protein chain.

In the 1930s -1950s, the first protein structures were solved by protein crystallography. These early structures suggested that a fixed three-dimensional structure might be generally required to mediate biological functions of proteins. When stating that proteins have just one uniquely defined configuration, Mirsky and Pauling did not recognize that Fisher's work would have supported their thesis with his 'Lock and Key' model (1894). These publications solidified the central dogma of molecular biology in that the sequence determines the structure which, in turn, determines the function of proteins. In 1950, Karush wrote about 'Configurational Adaptability' contradicting all the assumptions and research in the 19th century. He was convinced that proteins have more than one configuration at the same energy level and can choose one when binding to other substrates. In the 1960s, Levinthal's paradox suggested that the systematic conformational search of a long polypeptide is unlikely to yield a single folded protein structure on biologically relevant timescales (i.e. seconds to minutes). Curiously, for many (small) proteins or protein domains, relatively rapid and efficient refolding can be observed in vitro. As stated in Anfinsen's Dogma from 1973, the fixed 3D structure of these proteins is uniquely encoded in its primary structure (the amino acid sequence), is kinetically accessible and stable under a range of (near) physiological conditions, and can therefore be considered as the native state of such "ordered" proteins.

During the subsequent decades, however, many large protein regions could not be assigned in x-ray datasets, indicating that they occupy multiple positions which average out in electron density maps. The lack of a fixed, unique positions relative to the crystal lattice suggested that these regions were "disordered". Additional techniques for determining protein structures, such as NMR, demonstrated the presence of large flexible linkers and termini in many solved structural ensembles. It is now generally accepted that proteins exist as an ensemble of similar structures with some regions more constrained than others. Intrinsically Unstructured Proteins (IUPs) occupy the extreme end of this spectrum of flexibility, whereas IDPs also include proteins of considerable local structure tendency or flexible multidomain assemblies.These highly dynamic disordered regions of proteins have subsequently been linked to functionally important phenomena such as allosteric regulation and enzyme catalysis.

In the 2000s, bioinformatic predictions of intrinsic disorder in proteins indicated that intrinsic disorder is more common in sequenced/predicted proteomes than in known structures in the protein database. Based on DISOPRED2 prediction, long (>30 residue) disordered segments occur in 2.0% of archaean, 4.2% of eubacterial and 33.0% of eukaryotic proteins. In 2001, Dunker published his paper 'Intrinsically Disordered Proteins' questioning whether the newly found information was ignored for 50 years.

In the 2010s it became clear that IDPs are highly abundant among disease-related proteins.

Biological Roles

Many disordered proteins have the binding affinity with their receptors regulated by post-translational modification, thus it has been proposed that the flexibility of disordered proteins facilitates the different conformational requirements for binding the modifying enzymes as well as their receptors. Intrinsic disorder is particularly enriched in proteins implicated in cell signaling, transcription and chromatin remodeling functions.

Flexible Linkers

Disordered regions are often found as flexible linkers (or loops) connecting two globular or transmembrane domains. Linker sequences vary greatly in length and amino acid sequence, but are similar in amino acid composition (rich in polar uncharged amino acids). Flexible linkers allow the connecting domains to freely twist and rotate through space to recruit their binding partners or for those binding partners to induce larger scale interdomain conformation changes via protein domain dynamics.

Linear Motifs

Linear motifs are short disordered segments of proteins that mediate functional interactions with other proteins or other biomolecules (RNA, DNA, sugars etc.). Many roles or linear motifs are associated with cell regulation, for instance in control of cell shape, subcellular localisation of individual proteins and regulated protein turnover. Often, post-translational modifications such as phosphorylation tune the affinity (not rarely by several orders of magnitude) of individual linear motifs for specific interactions. Relatively rapid evolution and a relatively small number of structural restraints for establishing novel (low-affinity) interfaces make it particularly challenging to detect linear motifs but their widespread biological roles and the fact that many viruses mimick/hijack linear motifs to efficiently recode infected cells underlines the timely urgency of research on this very challenging and exciting topic. Unlike globular proteins IDPs do not have spatially-disposed active pockets. Nevertheless, in 80% of IDPs (~3 dozens) subjected to detailed structural characterization by NMR there are linear motifs termed PreSMos (pre-structured motifs) that are transient secondary structural elements primed for target recognition. In several cases it has been demonstrated that these transient structures become full and stable secondary structures, e.g., helices, upon target binding. Hence, PreSMos are the putative active sites in IDPs.

Coupled Folding and Binding

Many unstructured proteins undergo transitions to more ordered states upon binding to their targets (e.g. Molecular Recognition Features (MoRFs)). The coupled folding and binding may be local, involving only a few interacting residues, or it might involve an entire protein domain. It was recently shown that the coupled folding and binding allows the burial of a large surface area that would be possible only for fully structured proteins if they were much larger. Moreover, certain disordered regions might serve as "molecular switches" in regulating certain biological function by switching to ordered conformation upon molecular recognition like small molecule-binding, DNA/RNA binding, ion interactions etc.

The ability of disordered proteins to bind, and thus to exert a function, shows that stability is not a required condition. Many short functional sites, for example Short Linear Motifs are over-represented in disordered proteins.

Disorder in the Bound State (Fuzzy Complexes)

Intrinsically disordered proteins can retain their conformational freedom even when they bind specifically to other proteins. The structural disorder in bound state can be static or dynamic. In fuzzy complexes structural multiplicity is required for function and the manipulation of the bound disordered region changes activity. The conformational ensemble of the complex is modulated via post-translational modifications or protein interactions. Specificity of DNA binding proteins often depends on the length of fuzzy regions, which is varied by alternative splicing.

Structural Aspects

Intrinsically disordered proteins adapt many different structures in vivo according to the cell's conditions, creating a structural or conformational ensemble.

Therefore their structures are strongly function-related. However, only few proteins are fully disordered in their native state. Disorder is mostly found in intrinsically disordered regions (IDRs) within an otherwise well-structured protein. The term intrinsically disordered protein (IDP) therefore includes proteins that contain IDRs as well as fully disordered proteins.

The existence and kind of protein disorder is encoded in its amino acid sequence. In general, IDPs are characterized by a low content of bulky hydrophobic amino acids and a high proportion of polar and charged amino acids, usually referred to as low hydrophobicity. This property leads to good interactions with water. Furthermore high net charges promote disorder because of electrostatic repulsion resulting from equally charged residues. Thus disordered sequences cannot sufficiently bury a hydrophobic core to fold into stable globular proteins. In some cases, hydrophobic clusters in disordered sequences provide the clues for identifying the regions that undergo coupled folding and binding (refer to biological roles). Many disordered proteins reveal regions without any regular secondary structure These regions can be termed as flexible, compared to structured loops. While the latter are rigid and contain only one set of Ramachandran angles, IDPs involve multiple sets of angles. The term flexibility is also used for well-structured proteins, but describes a different phenomenon in the context of disordered proteins. Flexibility in structured proteins is bound to an equilibrium state, while it is not so in IDPs. Many disordered proteins also reveal low complexity sequences, i.e. sequences with over-representation of a few residues. While low complexity sequences are a strong indication of disorder, the reverse is not necessarily true, that is, not all disordered proteins have low complexity sequences. Disordered proteins have a low content of predicted secondary structure.

Experimental Validation

Intrinsically unfolded proteins, once purified, can be identified by various experimental methods. The primary method to obtain information on disordered regions of a protein is NMR spectroscopy. The lack of electron density in X-ray crystallographic studies may also be a sign of disorder.

Folded proteins have a high density (partial specific volume of 0.72-0.74 mL/g) and commensurately small radius of gyration. Hence, unfolded proteins can be detected by methods that are sensitive to molecular size, density or hydrodynamic drag, such as size exclusion chromatography, analytical ultracentrifugation, Small angle X-ray scattering (SAXS), and measurements of the diffusion constant. Unfolded proteins are also characterized by their lack of secondary structure, as assessed by far-UV (170-250 nm) circular dichroism (esp. a pronounced minimum at ~200 nm) or infrared spectroscopy. Unfolded proteins also have exposed backbone peptide groups exposed to solvent, so that they are readily cleaved by proteases, undergo rapid hydrogen-deuterium exchange and exhibit a small dispersion (<1 ppm) in their 1H amide chemical shifts as measured by NMR. (Folded proteins typically show dispersions as large as 5 ppm for the amide protons.) Recently, new methods including Fast parallel proteolysis (FASTpp) have been introduced, which allow to determine the fraction folded/disordered without the need for purification. Even subtle differences in the stability of missense mutations, protein partner binding and (self)polymerisation-induced folding of (e.g.) coiled-coils can be detected using FASTpp as recently demonstrated using the tropomyosin-troponin protein interaction. Fully unstructured protein regions can be experimentally validated by their hypersusceptibility to proteolysis using short digestion times and low protease concentrations.

Bulk methods to study IDP structure and dynamics include SAXS for ensemble shape information, NMR for atomistic ensemble refinement, Fluorescence for visualising molecular interactions and conformational transitions, x-ray crystallography to highlight more mobile regions in otherwise rigid protein crystals, cryo-EM to reveal less fixed parts of proteins, light scattering to monitor size distributions of IDPs or their aggregation kinetics, Circular Dichroism to monitor secondary structure of IDPs.

Single-molecule methods to study IDPs include spFRET to study conformational flexibility of IDPs and the kinetics of structural transitions, optical tweezers for high-resolution insights into the ensembles of IDPs and their oligomers or aggregates, nanopores to reveal global shape distributions of IDPs, magnetic tweezers to study structural transitions for long times at low forces, high-speed AFM to visualise the spatio-temporal flexibility of IDPs directly.

Disorder Prediction

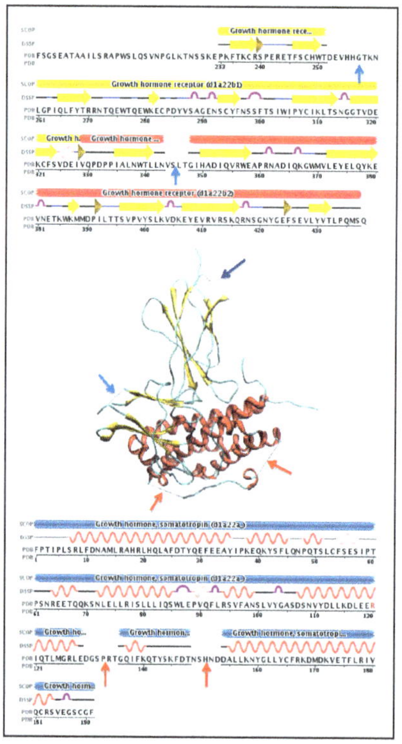

REMARK465 - missing electron densities in X-ray structure representing protein disorder (PDB: 1a22, human growth hormone bound to receptor). Compilation of screenshots from PDB database and molecule representation via VMD. Blue and red arrows point to missing residues on receptor and growth hormone, respectively.

Disorder prediction algorithms can predict Intrinsic Disorder (ID) propensity with high accuracy (approaching around 80%) based on primary sequence composition, similarity to unassigned segments in protein x-ray datasets, flexible regions in NMR studies and physico-chemical properties of amino acids.

Distinguishing IDPs from Well-structured Proteins

Separating disordered from ordered proteins is essential for disorder prediction. One of the first steps to find a factor that distinguishes IDPs from non-IDPs is to specify biases within the amino acid composition. The following hydrophilic, charged amino acids A, R, G, Q, S, P, E and K have been characterized as disorder-promoting amino acids, while order-promoting amino acids W, C, F, I, Y, V, L, and N are hydrophobic and uncharged. The remaining amino acids H, M, T and D are ambiguous, found in both ordered and unstructured regions. This information is the basis of most sequence-based predictors. Regions with little to no secondary structure, also known as NORS (NO Regular Secondary structure) regions, and low-complexity regions can easily be detected. However, not all disordered proteins contain such low complexity sequences.

Prediction Methods

Determining disordered regions from biochemical methods is very costly and time-consuming. Due to the variable nature of IDPs, only certain aspects of their structure can be detected, so that a full characterization requires a large number of different methods and experiments. This further increases the expense of IDP determination. In order to overcome this obstacle, computer-based methods are created for predicting protein structure and function. It is one of the main goals of bioinformatics to derive knowledge by prediction. Predictors for IDP function are also being developed, but mainly use structural information such as linear motif sites. There are different approaches for predicting IDP structure, such as neural networks or matrix calculations, based on different structural and/or biophysical properties.

Many computational methods exploit sequence information to predict whether a protein is disordered. Notable examples of such software include IUPRED and Disopred. Different methods may use different definitions of disorder. Meta-predictors show a new concept, combining different primary predictors to create a more competent and exact predictor.

Due to the different approaches of predicting disordered proteins, estimating their relative accuracy is fairly difficult. For example, neural networks are often trained on different datasets. The disorder prediction category is a part of biannual CASP experiment that is designed to test methods according accuracy in finding regions with missing 3D structure (marked in PDB files as REMARK465, missing electron densities in X-ray structures).

Disorder and Disease

Intrinsically unstructured proteins have been implicated in a number of diseases. Aggregation of misfolded proteins is the cause of many synucleinopathies and toxicity as those proteins start binding to each other randomly and can lead to cancer or cardiovascular diseases. Thereby, misfolding can happen spontaneously because millions of copies of proteins are made during the lifetime of an organism. The aggregation of the intrinsically unstructured protein α-Synuclein is thought to be responsible. The structural flexibility of this protein together with its susceptibility to modification in the cell leads to misfolding and aggregation. Genetics, oxidative and nitrative stress as well as mitochondrial impairment impact the structural flexibility of the unstructured α-Synuclein protein and associated disease mechanisms. Many key oncogenes have large intrinsically unstructured regions, for example p53 and BRCA1. These regions of the proteins are responsible for mediating many of their interactions. Taking the cell's native defense mechanisms as a model drugs can be developed, trying to block the place of noxious substrates and inhibiting them, and thus counteracting the disease.

Computer Simulations

Structural and dynamical properties of intrinsically unstructured proteins are being studied by molecular dynamics simulations. Findings from these simulations suggest a highly flexible conformational ensemble of intrinsically disordered proteins at different temperatures which is related to the presence of low free energy barriers.

Effects of confinement have also recently been addressed. For example, these studies suggest that confinement tends to increase the population of turn structures with respect to the population of coils and β-hairpins.

Moreover, various protocols and methods of analyzing IDPs, such as studies based on quantitative analysis of GC content in genes and their respective chromosomal bands, have been used to understand functional IDP segments.

Pioneering IDP Research Labs

In the last ten years, a great number of laboratories have investigated protein disorders using both experimental (e.g. SAXS-NMR, single-molecule fluorescence) and computational (analysis of protein structure) techniques.

Blood Proteins

Blood proteins, also termed plasma proteins or serum proteins, are proteins present in blood plasma. They serve many different functions, including transport of lipids, hormones, vitamins and minerals in the circulatory system and the regulation of acellular activity and functioning of the immune system. Other blood proteins act as enzymes, complement components, protease inhibitors or kinin precursors. Contrary to popular belief, hemoglobin is not a blood protein, as it is carried within red blood cells, rather than in the blood serum.

Serum albumin accounts for 55% of blood proteins, and is a major contributor to maintaining the osmotic pressure of plasma to assist in the transport of lipids and steroid hormones. Globulins make up 38% of blood proteins and transport ions, hormones, and lipids assisting in immune function. Fibrinogen comprises 7% of blood proteins; conversion of fibrinogen to insoluble fibrin is essential for blood clotting. The remainder of the plasma proteins (1%) are regulatory proteins, such as enzymes, proenzymes, and hormones. All blood proteins are synthesized in liver except for the gamma globulins.

Separating serum proteins by electrophoresis is a valuable diagnostic tool as well as a way to monitor clinical progress. Current research regarding blood plasma proteins is centered on performing proteomics analyses of serum/plasma in the search for biomarkers. These efforts started with two-dimensional gel electrophoresis efforts in the 1970s and in more recent times this research has been performed using LC-tandem MS based proteomics. The normal laboratory value of serum total protein is around 7 g/dL.

Families of blood proteins:

Blood protein	Normal level	%	Function
Albumins	3.5-5.0 g/dl	55%	maintain colloid osmotic pressure; create oncotic pressure and transport insoluble molecules
Globulins	2.0-2.5 g/dl	38%	participate in immune system
Fibrinogen	0.2-0.45 g/dl	7%	Blood coagulation
Regulatory proteins		<1%	Regulation of gene expression
Clotting factors		<1%	Conversion of fibrinogen into fibrin

Examples of specific blood proteins:

- Prealbumin (transthyretin)
- Alpha 1 antitrypsin (neutralizes trypsin that has leaked from the digestive system)
- Alpha 1 acid glycoprotein
- Alpha 1 fetoprotein
- Alpha2-macroglobulin
- Gamma globulins
- Beta-2 microglobulin
- Haptoglobin
- Ceruloplasmin
- Complement component 3
- Complement component 4
- C-reactive protein (CRP)
- Lipoproteins (chylomicrons, VLDL, LDL, HDL)
- Transferrin
- Prothrombin
- MBL or MBP

Globular Protein

Globular proteins or spheroproteins are spherical ("globe-like") proteins and are one of the common protein types (the others being fibrous, disordered and membrane proteins). Globular proteins are somewhat water-soluble (forming colloids in water), unlike the fibrous or membrane proteins. There are multiple fold classes of globular proteins, since there are many different architectures that can fold into a roughly spherical shape.

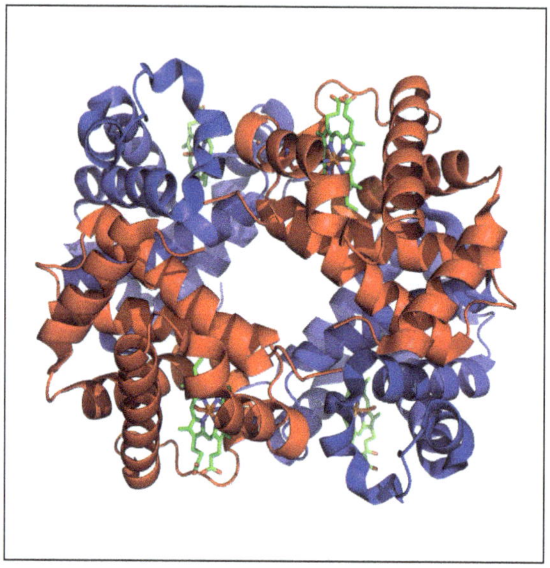

3-dimensional structure of hemoglobin, a globular protein.

The term globin can refer more specifically to proteins including the globin fold.

Globular Structure and Solubility

The term globular protein is quite old (dating probably from the 19th century) and is now somewhat archaic given the hundreds of thousands of proteins and more elegant and descriptive structural motif vocabulary. The globular nature of these proteins can be determined without the means of modern techniques, but only by using ultracentrifuges or dynamic light scattering techniques.

The spherical structure is induced by the protein's tertiary structure. The molecule's apolar (hydrophobic) amino acids are bounded towards the molecule's interior whereas polar (hydrophilic) amino acids are bound outwards, allowing dipole-dipole interactions with the solvent, which explains the molecule's solubility.

Globular proteins are only marginally stable because the free energy released when the protein folded into its native conformation is relatively small. This is because protein folding requires entropic cost. As a primary sequence of a polypeptide chain can form numerous conformations, native globular structure restricts its conformation to a few only. It results in a decrease in randomness, although non-covalent interactions such as hydrophobic interactions stabilize the structure.

Although it is still unknown how proteins fold up naturally, new evidence has helped advance understanding. Part of the protein folding problem is that several non-covalent, weak interactions are formed, such as hydrogen bonds and Van der Waals interactions. Via several techniques, the mechanism of protein folding is currently being studied. Even in the protein's denatured state, it can be folded into the correct structure.

Globular proteins seem to have two mechanisms for protein folding, either the diffusion-collision model or nucleation condensation model, although recent findings have shown globular proteins, such as PTP-BL PDZ2, that fold with characteristic features of both models. These new findings have shown that the transition states of proteins may affect the way they fold. The folding of globular proteins has also recently been connected to treatment of diseases, and anti-cancer ligands have been developed which bind to the folded but not the natural protein. These studies have shown that the folding of globular proteins affects its function.

By the second law of thermodynamics, the free energy difference between unfolded and folded states is contributed by enthalpy and entropy changes. As the free energy difference in a globular protein that results from folding into its native conformation is small, it is marginally stable, thus providing a rapid turnover rate and effective control of protein degradation and synthesis.

Role

Unlike fibrous proteins which only play a structural function, globular proteins can act as:

- Enzymes, by catalyzing organic reactions taking place in the organism in mild conditions and with a great specificity. Different esterases fulfill this role.
- Messengers, by transmitting messages to regulate biological processes. This function is done by hormones, i.e. insulin etc.
- Transporters of other molecules through membranes.
- Stocks of amino acids.
- Regulatory roles are also performed by globular proteins rather than fibrous proteins.
- Structural proteins, e.g., actin and tubulin, which are globular and soluble as monomers, but polymerize to form long, stiff fibers.

Members

Among the most known globular proteins is hemoglobin, a member of the globin protein family. Other globular proteins are the immunoglobulins (IgA, IgD, IgE, IgG and IgM), and alpha, beta and gamma globulins. Nearly all enzymes with major metabolic functions are globular in shape, as well as many signal transduction proteins.

Albumins are also globular proteins, although, unlike all of the other globular proteins, they are completely soluble in water.

Chapter 5

Common Techniques used in Proteomics

There are many methods and techniques used to investigate protein-protein interactions. A few of them are affinity chromatography, ChIP-sequencing, phosphotyrosine-binding domain, LIM domain, sterile alpha motif, FERM domain, protein subunit, RNA-binding protein, Pleckstrin homology domain, etc. The topics elaborated in this chapter will help in gaining a better perspective of this vast field.

Affinity Chromatography

Affinity chromatography is a method of separating biochemical mixtures based on a highly specific interaction such as that between antigen and antibody, enzyme and substrate, or receptor and ligand.

Uses

Affinity chromatography can be used to:

- Purify and concentrate a substance from a mixture into a buffering solution.
- Reduce the amount of a substance in a mixture.
- Discern what biological compounds bind to a particular substance.
- Purify and concentrate an enzyme solution.

Principle

The stationary phase is typically a gel matrix, often of agarose; a linear sugar molecule derived from algae. Usually the starting point is an undefined heterogeneous group of molecules in solution, such as a cell lysate, growth medium or blood serum. The molecule of interest will have a well known and defined property, and can be exploited during the affinity purification process. The process itself can be thought of as an entrapment, with the target molecule becoming trapped on a solid or stationary phase or medium. The other molecules in the mobile phase will not become trapped as they do not possess this property. The stationary phase can then be removed from the mixture, washed and the target molecule released from the entrapment in a process known as elution. Possibly the most common use of affinity chromatography is for the purification of recombinant proteins.

Batch and Column Setups

Binding to the solid phase may be achieved by column chromatography whereby the solid medium is packed onto a column, the initial mixture run through the column to allow setting, a wash buffer run through the column and the elution buffer subsequently applied to the column and collected. These steps are usually done at ambient pressure. Alternatively, binding may be achieved using a batch treatment, for example, by adding the initial mixture to the solid phase in a vessel, mixing, separating the solid phase, removing the liquid phase, washing, re-centrifuging, adding the elution buffer, re-centrifuging and removing the eluate.

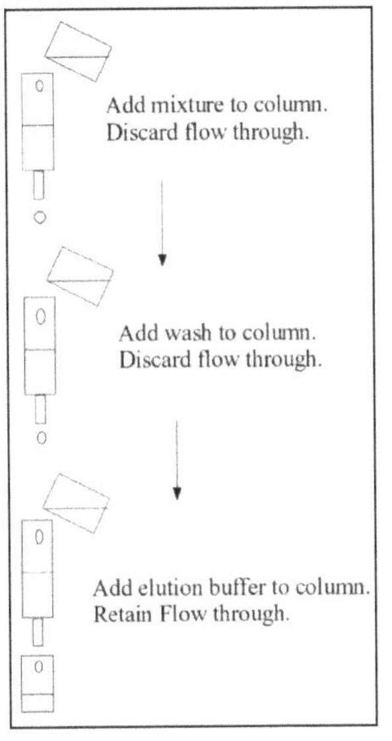

Column chromatography.

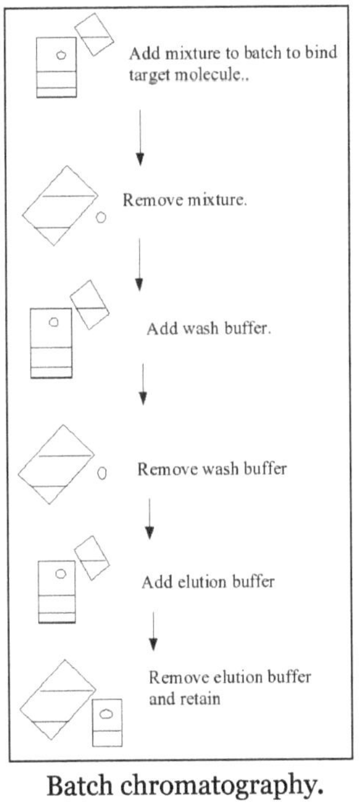

Batch chromatography.

Sometimes a hybrid method is employed such that the binding is done by the batch method, but the solid phase with the target molecule bound is packed onto a column and washing and elution are done on the column.

A third method, expanded bed adsorption, which combines the advantages of the two methods mentioned above, has also been developed. The solid phase particles are placed in a column where liquid phase is pumped in from the bottom and exits at the top. The gravity of the particles ensure that the solid phase does not exit the column with the liquid phase.

Affinity columns can be eluted by changing salt concentrations, pH, pI, charge and ionic strength directly or through a gradient to resolve the particles of interest.

More recently, setups employing more than one column in series have been developed. The advantage compared to single column setups is that the resin material can be fully loaded, since non-binding product is directly passed on to a consecutive column with fresh column material. The resin costs per amount of produced product can thus be drastically be reduced. Since one column can always be eluted and regenerated while the other column is loaded, already two columns are sufficient to make full use of the advantages. Additional columns can give additional flexibility for elution and regeneration times, at the cost of additional equipment and resin costs.

Specific Uses

Affinity chromatography can be used in a number of applications, including nucleic acid purification, protein purification from cell free extracts, and purification from blood.

Various Affinity Media

Many different affinity media exist for a variety of possible uses. Briefly, they are (generalized):

- Activated/Functionalized – Works as a functional spacer, support matrix, and eliminates handling of toxic reagents.
- Amino Acid – Used with a variety of serum proteins, proteins, peptides, and enzymes, as well as rRNA and dsDNA.
- Avidin Biotin – Used in the purification process of biotin/avidin and their derivatives.
- Carbohydrate Bonding – Most often used with glycoproteins or any other carbohydrate-containing substance.
- Carbohydrate – Used with lectins, glycoproteins, or any other carbohydrate metabolite protein.
- Dye Ligand – This media is nonspecific, but mimics biological substrates and proteins.
- Glutathione – Useful for separation of GST tagged recombinant proteins.
- Heparin – This media is a generalized affinity ligand, and it is most useful for separation of plasma coagulation proteins, along with nucleic acid enzymes and lipases.
- Hydrophobic Interaction – Most commonly used to target free carboxyl groups and proteins.
- Immunoaffinity – Detailed below, this method utilizes antigens' and antibodies' high specificity to separate.

- Immobilized Metal Affinity Chromatography – Detailed further below, this method uses interactions between metal ions and proteins (usually specially tagged) to separate.
- Nucleotide/Coenzyme – Works to separate dehydrogenases, kinases, and transaminases.
- Nucleic Acid – Functions to trap mRNA, DNA, rRNA, and other nucleic acids/oligonucleotides.
- Protein A/G – This method is used to purify immunoglobulins.
- Speciality – Designed for a specific class or type of protein/coenzyme, this type of media will only work to separate a specific protein or coenzyme.

Immunoaffinity

Another use for the procedure is the affinity purification of antibodies from blood serum. If serum is known to contain antibodies against a specific antigen (for example if the serum comes from an organism immunized against the antigen concerned) then it can be used for the affinity purification of that antigen. This is also known as Immunoaffinity Chromatography. For example if an organism is immunised against a GST-fusion protein it will produce antibodies against the fusion-protein, and possibly antibodies against the GST tag as well. The protein can then be covalently coupled to a solid support such as agarose and used as an affinity ligand in purifications of antibody from immune serum.

For thoroughness the GST protein and the GST-fusion protein can each be coupled separately. The serum is initially allowed to bind to the GST affinity matrix. This will remove antibodies against the GST part of the fusion protein. The serum is then separated from the solid support and allowed to bind to the GST-fusion protein matrix. This allows any antibodies that recognize the antigen to be captured on the solid support. Elution of the antibodies of interest is most often achieved using a low pH buffer such as glycine pH 2.8. The eluate is collected into a neutral tris or phosphate buffer, to neutralize the low pH elution buffer and halt any degradation of the antibody's activity. This is a nice example as affinity purification is used to purify the initial GST-fusion protein, to remove the undesirable anti-GST antibodies from the serum and to purify the target antibody.

A simplified strategy is often employed to purify antibodies generated against peptide antigens. When the peptide antigens are produced synthetically, a terminal cysteine residue is added at either the N- or C-terminus of the peptide. This cysteine residue contains a sulfhydryl functional group which allows the peptide to be easily conjugated to a carrier protein (e.g. Keyhole limpet hemocyanin (KLH)). The same cysteine-containing peptide is also immobilized onto an agarose resin through the cysteine residue and is then used to purify the antibody.

Most monoclonal antibodies have been purified using affinity chromatography based on immunoglobulin-specific Protein A or Protein G, derived from bacteria.

Immobilized Metal ion Affinity Chromatography

Immobilized metal ion affinity chromatography (IMAC) is based on the specific coordinate covalent bond of amino acids, particularly histidine, to metals. This technique works by allowing proteins with an affinity for metal ions to be retained in a column containing immobilized metal ions, such as cobalt, nickel, copper for the purification of histidine containing proteins or peptides, iron, zinc or gallium for the purification of phosphorylated proteins or peptides. Many naturally occurring proteins do not have an affinity for metal ions, therefore recombinant DNA technology can be used to introduce such a protein tag into the relevant gene. Methods used to elute the protein of interest include changing the pH, or adding a competitive molecule, such as imidazole.

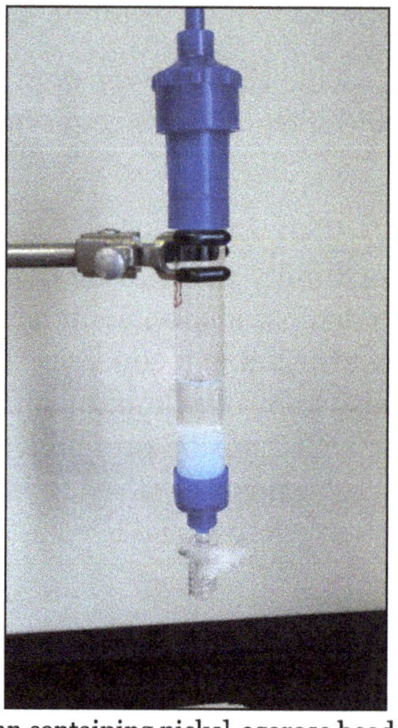

A chromatography column containing nickel-agarose beads used for purification of proteins with histidine tags.

Recombinant Proteins

Possibly the most common use of affinity chromatography is for the purification of recombinant proteins. Proteins with a known affinity are protein tagged in order to aid their purification. The protein may have been genetically modified so as to allow it to be selected for affinity binding; this is known as a fusion protein. Tags include glutathione-S-transferase (GST), hexahistidine (His), and maltose binding protein (MBP). Histidine tags have an affinity for nickel or cobalt ions which have been immobilized by forming coordinate covalent bonds with a chelator incorporated in the stationary phase. For elution, an excess amount of a compound able to act as a metal ion ligand, such as imidazole, is used. GST has an affinity for glutathione which is commercially available immobilized as glutathione agarose. During elution, excess glutathione is used to displace the tagged protein.

Lectins

Lectin affinity chromatography is a form of affinity chromatography where lectins are used to separate components within the sample. Lectins, such as Concanavalin A are proteins which can bind specific carbohydrate (sugar) molecules. The most common application is to separate glycoproteins from non-glycosylated proteins, or one glycoform from another glycoform.

ChIP-Sequencing

ChIP-sequencing, also known as ChIP-seq, is a method used to analyze protein interactions with DNA. ChIP-seq combines chromatin immunoprecipitation (ChIP) with massively parallel DNA sequencing to identify the binding sites of DNA-associated proteins. It can be used to map global binding sites precisely for any protein of interest. Previously, ChIP-on-chip was the most common technique utilized to study these protein–DNA relations.

Uses

ChIP-seq is used primarily to determine how transcription factors and other chromatin-associated proteins influence phenotype-affecting mechanisms. Determining how proteins interact with DNA to regulate gene expression is essential for fully understanding many biological processes and disease states. This epigenetic information is complementary to genotype and expression analysis. ChIP-seq technology is currently seen primarily as an alternative to ChIP-chip which requires a hybridization array. This necessarily introduces some bias, as an array is restricted to a fixed number of probes. Sequencing, by contrast, is thought to have less bias, although the sequencing bias of different sequencing technologies is not yet fully understood.

Specific DNA sites in direct physical interaction with transcription factors and other proteins can be isolated by chromatin immunoprecipitation. ChIP produces a library of target DNA sites bound to a protein of interest *in vivo*. Massively parallel sequence analyses are used in conjunction with whole-genome sequence databases to analyze the interaction pattern of any protein with DNA, or the pattern of any epigenetic chromatin modifications. This can be applied to the set of ChIP-able proteins and modifications, such as transcription factors, polymerases and transcriptional machinery, structural proteins, protein modifications, and DNA modifications. As an alternative to the dependence on specific antibodies, different methods have been developed to find the superset of all nucleosome-depleted or nucleosome-disrupted active regulatory regions in the genome, like DNase-Seq and FAIRE-Seq.

Workflow of ChIP-sequencing

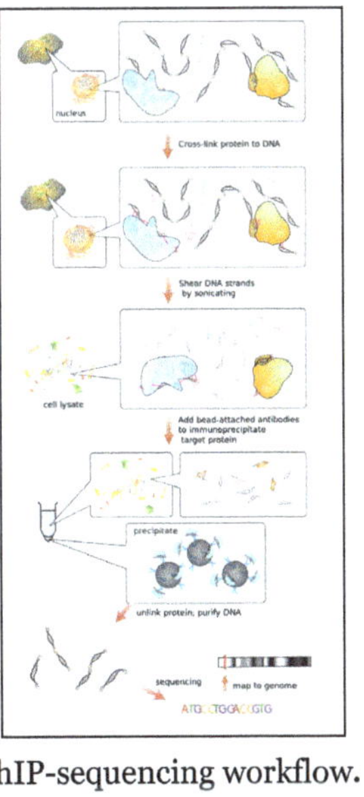

ChIP-sequencing workflow.

ChIP

ChIP is a powerful method to selectively enrich for DNA sequences bound by a particular protein in living cells. However, the widespread use of this method has been limited by the lack of a sufficiently robust method to identify all of the enriched DNA sequences. The ChIP process enriches specific crosslinked DNA-protein complexes using an antibody against the protein of interest. Oligonucleotide adaptors are then added to the small stretches of DNA that were bound to the protein of interest to enable mas-sively parallel sequencing.

Sequencing

After size selection, all the resulting ChIP-DNA fragments are sequenced simultaneously using a genome sequencer. A single sequencing run can scan for genome-wide associations with high resolution, meaning that features can be located precisely on the chromosomes. ChIP-chip, by contrast, requires large sets of tiling arrays for lower resolution.

There are many new sequencing methods used in this sequencing step. Some technologies that analyze the sequences can use cluster amplification of adapter-ligated ChIP DNA fragments on a solid flow cell substrate to create clusters of approximately 1000 clonal copies each. The resulting high density array of template clusters on the flow cell surface is sequenced by a Genome analyzing program. Each template cluster undergoes sequencing-by-synthesis in parallel using novel fluorescently labelled reversible terminator nucleotides. Templates are sequenced base-by-base during each read. Then, the data collection and analysis software aligns sample sequences to a known genomic sequence to identify the ChIP-DNA fragments.

Sensitivity

Sensitivity of this technology depends on the depth of the sequencing run (i.e. the number of mapped sequence tags), the size of the genome and the distribution of the target factor. The sequencing depth is directly correlated with cost. If abundant binders in large genomes have to be mapped with high sensitivity, costs are high as an enormously high number of sequence tags will be required. This is in contrast to ChIP-chip in which the costs are not correlated with sensitivity.

Unlike microarray-based ChIP methods, the precision of the ChIP-seq assay is not limited by the spacing of predetermined probes. By integrating a large number of short reads, highly precise binding site localization is obtained. Compared to ChIP-chip, ChIP-seq data can be used to locate the binding site within few tens of base pairs of the actual protein binding site. Tag densities at the binding sites are a good indicator of protein–DNA binding affinity, which makes it easier to quantify and compare binding affinities of a protein to different DNA sites.

Current Research

STAT1 DNA association: ChIP-seq was used to study STAT1 targets in HeLA S3 cells. The performance of ChIP-seq was then compared to the alternative protein–DNA interaction methods of ChIP-PCR and ChIP-chip.

Nucleosome Architecture of Promoters: Using ChIP-seq, it was determined that Yeast genes seem to have a minimal nucleosome-free promoter region of 150bp in which RNA polymerase can initiate transcription.

Transcription factor conservation: ChIP-seq was used to compare conservation of TFs in the forebrain and heart tissue in embryonic mice. The authors identified and validated the heart functionality of transcription enhancers, and determined that transcription enhancers for the heart are less conserved than those for the forebrain during the same developmental stage.

Genome-wide ChIP-seq: ChIP-sequencing was completed on the worm C. elegans to explore genome-wide binding sites of 22 transcription factors. Up to 20% of the annotated candidate genes were assigned to transcription factors. Several transcription factors were assigned to non-coding RNA regions and may be subject to developmental or environmental variables. The functions of some of the transcription factors were also identified. Some of the transcription factors regulate genes that control other transcription factors. These genes are not regulated by other factors. Most transcription factors serve as both targets and regulators of other factors, demonstrating a network of regulation.

Inferring regulatory network: ChIP-seq signal of Histone modification were shown to be more correlated with transcription factor motifs at promoters in comparison to RNA level. Hence author proposed that using histone modification ChIP-seq would provide more reliable inference of gene-regulatory networks in comparison to other methods based on expression.

ChIP-seq offers an alternative to ChIP-chip. STAT1 experimental ChIP-seq data have a high degree of similarity to results obtained by ChIP-chip for the same type of experiment, with >64% of peaks in shared genomic regions. Because the data are sequence reads, ChIP-seq offers a rapid analysis pipeline (as long as a high-quality genome sequence is available for read mapping, and the genome doesn't have repetitive content that confuses the mapping process) as well as the potential to detect mutations in binding-site sequences, which may directly support any observed changes in protein binding and gene regulation.

Computational Analysis

As with many high-throughput sequencing approaches, ChIP-seq generates extremely large data sets, for which appropriate computational analysis methods are required. To predict DNA-binding sites from ChIP-seq read count data, peak calling methods have been developed. The most popular method is MACS which empirically models the shift size of ChIP-Seq tags, and uses it to improve the spatial resolution of predicted binding sites.

Another relevant computational problem is Differential peak calling, which identifies significant differences in two ChIP-seq signals from distinct biological conditions. Differential peak callers segment two ChIP-seq signals and identify differential peaks using Hidden Markov Models. Examples for two-stage differential peak callers are ChIPDiff and ODIN.

Protein-Fragment Complementation Assay

A Protein-fragment complementation assay, or PCA, is a method for the identification of protein–protein interactions, especially in the field of proteomics. In the PCA, the proteins of interest ("bait" and "prey") are each covalently linked to incomplete fragments of a third protein (e.g.

DHFR, which acts as a "reporter"). Interaction between the bait and the prey proteins brings the fragments of the reporter protein in close enough proximity to allow them to form a functional reporter protein whose activity can be measured. This principle can be applied to many different reporter proteins and is also the basis for the yeast two-hybrid system, an archetypical PCA assay.

Split Protein Assays

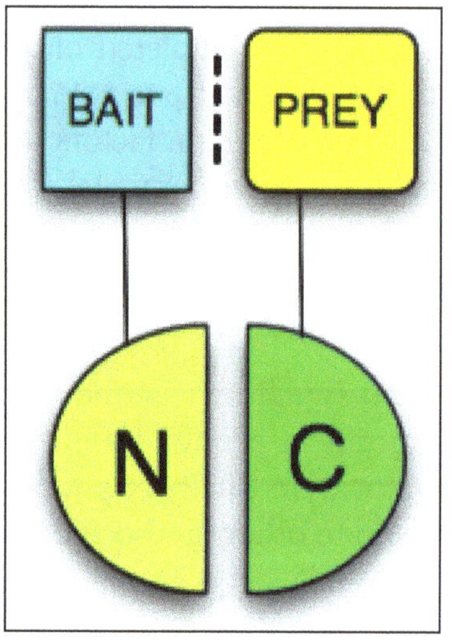

General principle of the protein complementation assay: a protein is split into two (N- and C-terminal) halves and reconstituted by two interacting proteins (here called "bait" and "prey" because a bait protein can be used to find an interacting prey protein). The activity of the reconstituted protein should be easily detectable, e.g. as in the Green Fluorescent Protein (GFP).

Any protein that can be split into two parts and reconstituted non-covalently may be used in a PCA. The two parts just have to be brought together by other interacting proteins fused to them (often called "bait" and "prey" because a bait protein can be used to find a prey protein). The protein that produces a detectable readout is called "reporter". Usually enzymes which confer resistance to antibiotics, such as Dihydrofolate reductase or Beta-lactamase, or proteins that give colorimetric or fluorescent signals are used as reporters. When fluorescent proteins are reconstituted the PCA is called Bimolecular fluorescence complementation assay. The following proteins have been used in split protein PCAs:

- Beta-lactamase.
- Dihydrofolate reductase (DHFR).
- Focal adhesion kinase (FAK).
- Gal4, a yeast transcription factor (as in the classical yeast two-hybrid system).
- GFP (split-GFP), e.g. EGFP (enhanced green fluorescent protein).
- Horseradish peroxidase.
- Infrared fluorescent protein IFP1.4, an engineered chromophore-binding domain (CBD) of a bacteriophytochrome from *Deinococcus radiodurans*.

- LacZ (beta-galactosidase).
- Luciferase, including ReBiL (recombinase enhanced bimolecular luciferase) and the commercial products NanoLuc and NanoBIT from Promega.
- TEV (Tobacco etch virus protease).
- Ubiquitin.

Protein Footprinting

Protein footprinting is a term used to refer to a method of biochemical analysis that investigates protein structure, assembly, and interactions within a larger macromolecular assembly. It was originally coined in reference to the use of limited proteolysis to investigate contact sites within a monoclonal antibody - protein antigen complex and a year later to examine the protection from hydroxyl radical cleavage conferred by a protein bound to DNA within a DNA-protein complex. In DNA footprinting the protein is envisioned to make an imprint (or footprint) at a particular point of interaction. This latter method was adapted through the direct treatment of proteins and their complexes with hydroxyl radicals.

Hydroxyl Radical Protein Footprinting

Time-resolved hydroxyl radical protein footprinting employing mass spectrometry analysis was developed in the late 1990s in synchrotron radiolysis studies. The same year, these authors reported on the use of an electrical discharge source to effect the oxidation of proteins on millisecond timescales as proteins pass from the electrosprayed solution into the mass spectrometer. These approaches have since been used to determine protein structures, protein folding, protein dynamics, and protein–protein interactions.

Unlike nucleic acids, proteins oxidize rather than cleave on these timescales. Analysis of the products by mass spectrometry reveals that proteins to are oxidized in a limited manner (some 10–30% of total protein) at a number of amino acid side chains across the proteins. The rate or level of oxidation at the reactive amino acid side chains (Met, Cys, Trp, Tyr, Phe, His, Pro and Leu) provides a measure of their accessibility to the bulk solvent. The mechanisms of side chain oxidation was explored by performing the radiolysis reactions in ^{18}O-labeled water.

Producing OH Radicals

A critical feature of these experiments is the need to expose proteins to hydroxyl radicals for limited timescales on the order of 1–50 ms inducing 10-30% oxidation of total protein. A further requirement is to generate hydroxyl radicals from the bulk solvent (i.e. water) (equations 1 and 2) not hydrogen peroxide which can remain to oxidize proteins even without other stimuli.

$$H_2O \rightarrow H_2O^{+\cdot} + e^- + H_2O^*$$

$$H_2O^{+\cdot} + H_2O \rightarrow H_3O^+ + OH^\cdot$$

Hydroxyl radicals can be produced in solution by an electrical discharge within a conventional atmospheric pressure electrospray ionization (ESI) source. When a high voltage difference (~8 keV) is held between an electrospray needle and a sampling orifice to the mass analyzer, radicals can be produced in solution at the electrospray needle tip. This method was the first employed to apply protein footprinting to the study of a protein complex.

Method

The exposure of proteins to a "white" X-ray beam of synchrotron light or an electrical discharge for tens of milliseconds provides sufficient oxidative modification to the surface amino acid side chains without damage to the protein structure. These products can be easily detected and quantified by mass spectrometry. By adjusting the time for radiolysis or which protein ions spend in the discharge source, a time-resolved approach is possible which is valuable for the study of protein dynamics.

Analysis

A computer program (PROXIMO) has also been written to help model protein complexes using data from the RP-MS/Protein footprinting approach. RP-MS/Protein footprinting studies of protein complexes can also employ computational approaches to assist with this modeling.

Applications

The application of ion mobility mass spectrometry has conclusively demonstrated that the conditions employed in RP-MS/Protein footprinting experiments do not alter the structure of proteins.

Other studies have extended the method to study early onset protein damage given the radical basis of the method and the significance of oxygen based radicals in the pathogenesis of many diseases including neurological disorders and even blindness.

Microscale Thermophoresis

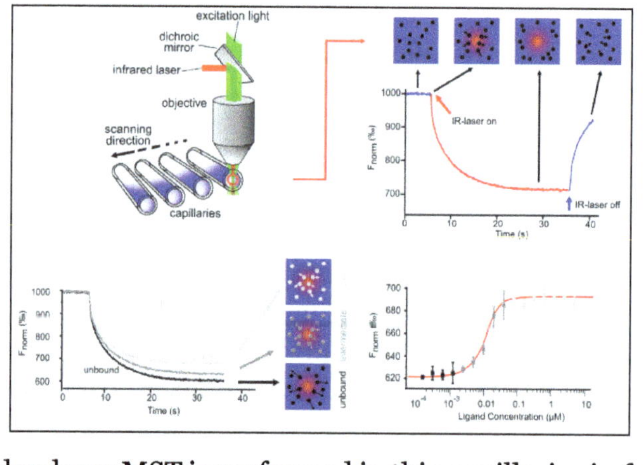

Principle of the MST Technology: MST is performed in thin capillaries in free solution thus providing close-to-native conditions (immobilization free in any buffer, even in complex bioliquids) and a maintenance free instrument. When performing a MST experiment, a microscopic temperature gradient is induced by an infrared laser, and the directed movement of molecules is detected and quantified. Thermophoresis, the movement of the molecule in the temperature gradient, depends on three parameters that typically change upon interaction. Thus, the thermophoresis signal is plotted against the ligand concentration to obtain a dose-response curve, from which the binding affinity can be deduced.

Microscale thermophoresis (MST) is a technology for the interaction analysis of biomolecules. Microscale thermophoresis is the directed movement of particles in a microscopic temperature gradient. Any change of the hydration shell of biomolecules due to changes in their structure/ conformation results in a relative change of the movement along the temperature gradient and is used to determine binding affinities. MST allows measurement of interactions directly in solution without the need of immobilization to a surface (immobilization-free technology).

Applications

Affinity:

- Between any kind of biomolecules including proteins, DNA, RNA, peptides, small molecules, fragments and ions.
- For interactions with high molecular weight complexes, large molecule assemblies, even with liposomes, vesicles, nanodiscs, nanoparticles and viruses.
- In any buffer, including serum and cell lysate.
- In competition experiments (for example with subtrate and inhibitors).

Stoichiometry

Thermodynamic parameters

Additional information:

- Sample property (homogeneity, aggregation, stability).
- Multiple binding sites, cooperativity.

Technology

MST is based on the directed movement of molecules along temperature gradients, an effect termed thermophoresis. A spatial temperature difference ΔT leads to a depletion of molecule concentration in the region of elevated temperature, quantified by the Soret coefficient S_T: $c_{hot}/c_{cold} = \exp(-S_T \Delta T)$.

Thermophoresis depends on the interface between molecule and solvent. Under constant buffer conditions, thermophoresis probes the size, charge and solvation entropy of the molecules. The thermophoresis of a fluorescently labeled molecule A typically differs significantly from the thermophoresis of a molecule-target complex AT due to size, charge and solvation entropy differences. This difference in the molecule's thermophoresis is used to quantify the binding in titration experiments under constant buffer conditions.

The thermophoretic movement of the fluorescently labelled molecule is measured by monitoring the fluorescence distribution F inside a capillary . The microscopic temperature gradient is generated by an IR-Laser, which is focused into the capillary and is strongly absorbed by water. The temperature of the aqueous solution in the laser spot is raised by up to $\Delta T=5$ K. Before the IR-Laser is switched on a homogeneous fluorescence distribution F_{cold} is observed inside the capillary. When the IR-Laser is switched on, two effects, separated by their time-scales, contribute to the new fluorescence distribution F_{hot}. The thermal relaxation time is fast and induces a binding-dependent drop in the fluorescence of the dye due to its local environmental-dependent response to the temperature jump. On the slower diffusive time scale (10 s), the molecules move from the locally heated region to the outer cold regions. The local concentration of molecules decreases in the heated region until it reaches a steady-state distribution.

While the mass diffusion D dictates the kinetics of depletion, S_T determines the steady-state concentration ratio $c_{hot}/c_{cold}=\exp(-S_T\,\Delta T) \approx 1-S_T\,\Delta T$ under a temperature increase ΔT. The normalized fluorescence $F_{norm}=F_{hot}/F_{cold}$ measures mainly this concentration ratio, in addition to the temperature jump $\partial F/\partial T$. In the linear approximation we find: $F_{norm}=1+(\partial F/\partial T-S_T)\Delta T$. Due to the linearity of the fluorescence intensity and the thermophoretic depletion, the normalized fluorescence from the unbound molecule $F_{norm}(A)$ and the bound complex $F_{norm}(AT)$ superpose linearly. By denoting x the fraction of molecules bound to targets, the changing fluorescence signal during the titration of target T is given by: $F_{norm}=(1-x)\,F_{norm}(A)+x\,F_{norm}(AT)$.

Quantitative binding parameters are obtained by using a serial dilution of the binding substrate. By plotting F_{norm} against the logarithm of the different concentrations of the dilution series, a sigmoidal binding curve is obtained. This binding curve can directly be fitted with the nonlinear solution of the law of mass action, with the dissociation constant K_D as result.

Protein Microarray

A protein microarray (or protein chip) is a high-throughput method used to track the interactions and activities of proteins, and to determine their function, and determining function on a large scale. Its main advantage lies in the fact that large numbers of proteins can be tracked in parallel. The chip consists of a support surface such as a glass slide, nitrocellulose membrane, bead, or microtitre plate, to which an array of capture proteins is bound. Probe molecules, typically labeled with a fluorescent dye, are added to the array. Any reaction between the probe and the immobilised protein emits a fluorescent signal that is read by a laser scanner. Protein microarrays are rapid, automated, economical, and highly sensitive, consuming small quantities of samples and reagents. The concept and methodology of protein microarrays was first introduced and illustrated in antibody microarrays (also referred to as antibody matrix) in 1983 in a scientific publication and a series of patents. The high-throughput technology behind the protein microarray was relatively easy to develop since it is based on the technology developed for DNA microarrays, which have become the most widely used microarrays.

Motivation for Development

Protein microarrays were developed due to the limitations of using DNA microarrays for determining gene expression levels in proteomics. The quantity of mRNA in the cell often doesn't reflect the expression levels of the proteins they correspond to. Since it is usually the protein, rather than the mRNA, that has the functional role in cell response, a novel approach was needed. Additionally post-translational modifications, which are often critical for determining protein function, are not visible on DNA microarrays. Protein microarrays replace traditional proteomics techniques such as 2D gel electrophoresis or chromatography, which were time consuming, labor-intensive and ill-suited for the analysis of low abundant proteins.

Making the Array

The proteins are arrayed onto a solid surface such as microscope slides, membranes, beads or microtitre plates. The function of this surface is to provide a support onto which proteins can be immobilized. It should demonstrate maximal binding properties, whilst maintaining the protein in its native conformation so that its binding ability is retained. Microscope slides made of glass or silicon are a popular choice since they are compatible with the easily obtained robotic arrayers and laser scanners that have been developed for DNA microarray technology. Nitrocellulose film slides are broadly accepted as the highest protein binding substrate for protein microarray applications.

The chosen solid surface is then covered with a coating that must serve the simultaneous functions of immobilising the protein, preventing its denaturation, orienting it in the appropriate direction so that its binding sites are accessible, and providing a hydrophilic environment in which the binding reaction can occur. In addition, it also needs to display minimal non-specific binding in order to minimize background noise in the detection systems. Furthermore, it needs to be compatible with different detection systems. Immobilising agents include layers of aluminium or gold, hydrophilic polymers, and polyacrylamide gels, or treatment with amines, aldehyde or epoxy. Thin-film technologies like physical vapour deposition (PVD) and chemical vapour deposition (CVD) are employed to apply the coating to the support surface.

An aqueous environment is essential at all stages of array manufacture and operation to prevent protein denaturation. Therefore, sample buffers contain a high percent of glycerol(to lower the freezing point), and the humidity of the manufacturing environment is carefully regulated. Microwells have the dual advantage of providing an aqueous environment while preventing cross-contamination between samples.

In the most common type of protein array, robots place large numbers of proteins or their ligands onto a coated solid support in a pre-defined pattern. This is known as robotic contact printing or robotic spotting. Another fabrication method is ink-jetting, a drop-on-demand, non-contact method of dispersing the protein polymers onto the solid surface in the desired pattern. Piezoelectric spotting is a similar method to ink-jet printing. The printhead moves across the array, and at each spot uses electric stimulation to deliver the protein molecules onto the surface via tiny jets. This is also a non-contact process. Photolithography is a fourth method of arraying the proteins onto the surface. Light is used in association with photomasks, opaque plates with holes or transparencies that allow light to shine through in a defined pattern. A series of chemical treatments then enables deposition of the protein in the desired pattern upon the material underneath the photomask.

The capture molecules arrayed on the solid surface may be antibodies, antigens, aptamers (nucleic acid-based ligands), affibodies (small molecules engineered to mimic monoclonal antibodies), or full length proteins. Sources of such proteins include cell-based expression systems for recombinant proteins, purification from natural sources, production in vitro by cell-free translation systems, and synthetic methods for peptides. Many of these methods can be automated for high throughput production but care must be taken to avoid conditions of synthesis or extraction that result in a denatured protein which, since it no longer recognizes its binding partner, renders the array useless.

Proteins are highly sensitive to changes in their microenvironment. This presents a challenge in maintaining protein arrays in a stable condition over extended periods of time. In situ methods involve on-chip synthesis of proteins as and when required, directly from the DNA using cell-free protein expression systems. Since DNA is a highly stable molecule it does not deteriorate over time and is therefore suited to long-term storage. This approach is also advantageous in that it circumvents the laborious and often costly processes of separate protein purification and DNA cloning, since proteins are made and immobilised simultaneously in a single step on the chip surface. Examples of In situ techniques are PISA (protein in situ array), NAPPA (nucleic acid programmable protein array) and DAPA (DNA array to protein array).

Types of Arrays

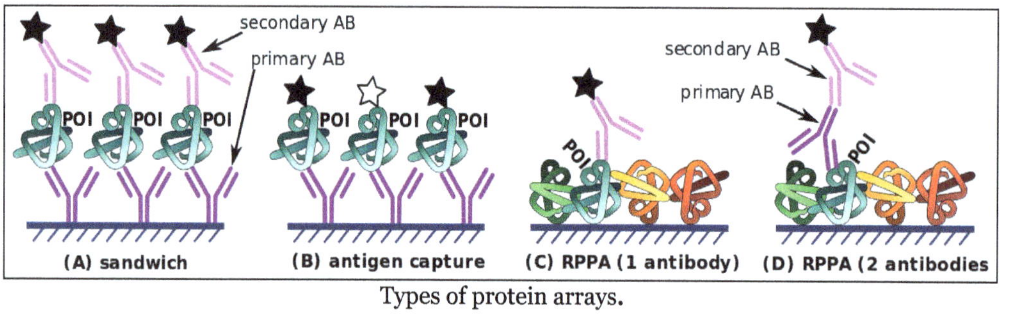

Types of protein arrays.

There are three types of protein microarrays that are currently used to study the biochemical activities of proteins.

Analytical microarrays are also known as capture arrays. In this technique, a library of antibodies, aptamers or affibodies is arrayed on the support surface. These are used as capture molecules since each binds specifically to a particular protein. The array is probed with a complex protein solution such as a cell lysate. Analysis of the resulting binding reactions using various detection systems can provide information about expression levels of particular proteins in the sample as well as measurements of binding affinities and specificities. This type of microarray is especially useful in comparing protein expression in different solutions. For instance the response of the cells to a particular factor can be identified by comparing the lysates of cells treated with specific substances or grown under certain conditions with the lysates of control cells. Another application is in the identification and profiling of diseased tissues.

Functional protein microarrays (also known as target protein arrays) are constructed by immobilising large numbers of purified proteins and are used to identify protein-protein, protein-DNA, protein-RNA, protein-phospholipid, and protein-small molecule interactions, to assay enzymatic activity and to detect antibodies and demonstrate their specificity. They differ from analytical arrays in that functional protein arrays are composed of arrays containing full-length functional proteins or protein domains. These protein chips are used to study the biochemical activities of the entire proteome in a single experiment.

Reverse phase protein microarray (RPPA) involve complex samples, such as tissue lysates. Cells are isolated from various tissues of interest and are lysed. The lysate is arrayed onto the microarray and probed with antibodies against the target protein of interest. These antibodies are typically detected with chemiluminescent, fluorescent or colorimetric assays. Reference peptides are printed on the slides to allow for protein quantification of the sample lysates. RPAs allow for the determination of the presence of altered proteins or other agents that may be the result of disease. Specifically, post-translational modifications, which are typically altered as a result of disease can be detected using RPAs.

Detection

Protein array detection methods must give a high signal and a low background. The most common and widely used method for detection is fluorescence labeling which is highly sensitive, safe and compatible with readily available microarray laser scanners. Other labels can be used, such as affinity, photochemical or radioisotope tags. These labels are attached to the probe itself and can interfere with the probe-target protein reaction. Therefore, a number of label free detection methods are available, such as surface plasmon resonance (SPR), carbon nanotubes, carbon nanowire sensors (where detection occurs via changes in conductance) and microelectromechanical system (MEMS) cantilevers. All these label free detection methods are relatively new and are not yet suitable for high-throughput protein interaction detection; however, they do offer much promise for the future.

Protein quantitation on nitrocellulose coated glass slides can use near-IR fluorescent detection. This limits interferences due to auto-fluorescence of the nitrocellulose at the UV wavelengths used for standard fluorescent detection probes.

Applications

There are five major areas where protein arrays are being applied: diagnostics, proteomics, protein functional analysis, antibody characterization, and treatment development.

Diagnostics involves the detection of antigens and antibodies in blood samples; the profiling of sera to discover new disease biomarkers; the monitoring of disease states and responses to therapy in personalized medicine; the monitoring of environment and food.

Proteomics pertains to protein expression profiling i.e. which proteins are expressed in the lysate of a particular cell.

Protein functional analysis is the identification of protein-protein interactions (e.g. identification of members of a protein complex), protein-phospholipid interactions, small molecule targets, enzymatic substrates (particularly the substrates of kinases) and receptor ligands.

Antibody characterization is characterizing cross-reactivity, specificity and mapping epitopes.

Treatment development involves the development of antigen-specific therapies for autoimmunity, cancer and allergies; the identification of small molecule targets that could potentially be used as new drugs.

Challenges

Despite the considerable investments made by several companies, proteins chips have yet to flood the market. Manufacturers have found that proteins are actually quite difficult to handle. A protein chip requires a lot more steps in its creation than does a DNA chip.

Challenges include: 1) finding a surface and a method of attachment that allows the proteins to maintain their secondary or tertiary structure and thus their biological activity and their interactions with other molecules, 2) producing an array with a long shelf life so that the proteins on the chip do not denature over a short time, 3) identifying and isolating antibodies or other capture molecules against every protein in the human genome, 4) quantifying the levels of bound protein while assuring sensitivity and avoiding background noise, 5) extracting the detected protein from the chip in order to further analyze it, 6) reducing non-specific binding by the capture agents, 7) the capacity of the chip must be sufficient to allow as complete a representation of the proteome to be visualized as possible; abundant proteins overwhelm the detection of less abundant proteins such as signaling molecules and receptors, which are generally of more therapeutic interest.

Two-Hybrid Screening

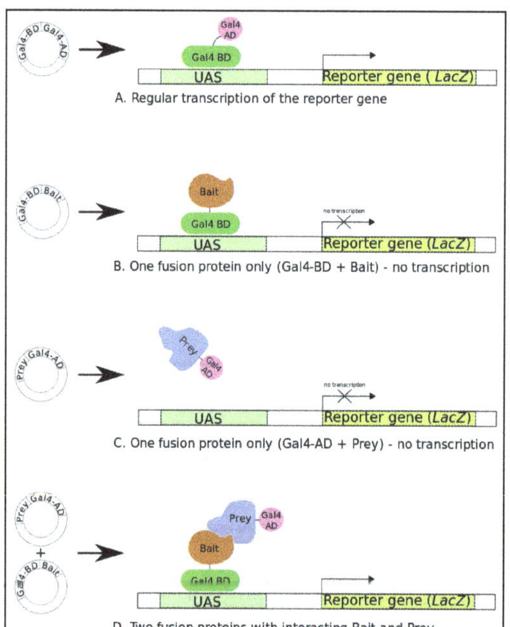

Overview of two-hybrid assay, checking for interactions between two proteins, called here *Bait* and *Prey*. A. The *Gal4* transcription factor gene produces a two-domain protein (*BD* and *AD*) essential for transcription of the reporter gene (*LacZ*). B,C. Two fusion proteins are prepared: *Gal4BD+Bait* and *Gal4AD+Prey*. Neither of them are usually sufficient to initiate transcription (of the reporter gene) alone.D. When both fusion proteins are produced and the Bait part of the first fusion protein interacts with the Prey part of the second, transcription of the reporter gene occurs.

Two-hybrid screening (also known as yeast two-hybrid system or Y2H) is a molecular biology technique used to discover protein–protein interactions (PPIs) and protein–DNA interactions by testing for physical interactions (such as binding) between two proteins or a single protein and a DNA molecule, respectively.

The premise behind the test is the activation of downstream reporter gene(s) by the binding of a transcription factor onto an upstream activating sequence (UAS). For two-hybrid screening, the transcription factor is split into two separate fragments, called the binding domain (BD) and activating domain (AD). The BD is the domain responsible for binding to the UAS and the AD is the domain responsible for the activation of transcription. The Y2H is thus a protein-fragment complementation assay.

History

Pioneered by Stanley Fields and Ok-Kyu Song in 1989, the technique was originally designed to detect protein–protein interactions using the GAL4 transcriptional activator of the yeast *Saccharomyces cerevisiae*. The GAL4 protein activated transcription of a protein involved in galactose utilization, which formed the basis of selection. Since then, the same principle has been adapted to describe many alternative methods, including some that detect protein–DNA interactions or DNA-DNA interactions, as well as methods that use *Escherichia coli* instead of yeast.

Basic Premise

The key to the two-hybrid screen is that in most eukaryotic transcription factors, the activating and binding domains are modular and can function in proximity to each other without direct binding. This means that even though the transcription factor is split into two fragments, it can still activate transcription when the two fragments are indirectly connected.

The most common screening approach is the yeast two-hybrid assay. This system often utilizes a genetically engineered strain of yeast in which the biosynthesis of certain nutrients (usually amino acids or nucleic acids) is lacking. When grown on media that lacks these nutrients, the yeast fail to survive. This mutant yeast strain can be made to incorporate foreign DNA in the form of plasmids. In yeast two-hybrid screening, separate bait and prey plasmids are simultaneously introduced into the mutant yeast strain.

Plasmids are engineered to produce a protein product in which the DNA-binding domain (BD) fragment is fused onto a protein while another plasmid is engineered to produce a protein product in which the activation domain (AD) fragment is fused onto another protein. The protein fused to the BD may be referred to as the bait protein, and is typically a known protein the investigator is using to identify new binding partners. The protein fused to the AD may be referred to as the prey protein and can be either a single known protein or a library of known or unknown proteins. In this context, a library may consist of a collection of protein-encoding sequences that represent all the proteins expressed in a particular organism or tissue, or may be generated by synthesising random DNA sequences. Regardless of the source, they are subsequently incorporated into the protein-encoding sequence of a plasmid, which is then transfected into the cells chosen for the screening method. This technique, when using a library, assumes that each cell is transfected with no more than a single plasmid and that, therefore, each cell ultimately expresses no more than a single member from the protein library.

If the bait and prey proteins interact (i.e., bind), then the AD and BD of the transcription factor are indirectly connected, bringing the AD in proximity to the transcription start site and transcription of reporter gene(s) can occur. If the two proteins do not interact, there is no transcription of the reporter gene. In this way, a successful interaction between the fused protein is linked to a change in the cell phenotype.

The challenge of separating cells that express proteins that happen to interact with their counterpart fusion proteins from those that do not, is addressed in the following section.

Fixed Domains

In any study, some of the protein domains, those under investigation, will be varied according to the goals of the study whereas other domains, those that are not themselves being investigated, will be kept constant. For example, in a two-hybrid study to select DNA-binding domains, the DNA-binding domain, BD, will be varied while the two interacting proteins, the bait and prey, must be kept constant to maintain a strong binding between the BD and AD. There are a number of domains from which to choose the BD, bait and prey and AD, if these are to remain constant. In protein–protein interaction investigations, the BD may be chosen from any of many strong DNA-binding domains such as Zif268. A frequent choice of bait and prey domains are residues 263–352 of yeast Gal11P with a N342V mutation and residues 58–97 of yeast Gal4, respectively. These domains can be used in both yeast- and bacterial-based selection techniques and are known to bind together strongly.

The AD chosen must be able to activate transcription of the reporter gene, using the cell's own transcription machinery. Thus, the variety of ADs available for use in yeast-based techniques may not be suited to use in their bacterial-based analogues. The herpes simplex virus-derived AD, VP16 and yeast Gal4 AD have been used with success in yeast whilst a portion of the α-subunit of *E. coli* RNA polymerase has been utilised in *E. coli*-based methods.

Whilst powerfully activating domains may allow greater sensitivity towards weaker interactions, conversely, a weaker AD may provide greater stringency.

Construction of Expression Plasmids

A number of engineered genetic sequences must be incorporated into the host cell to perform two-hybrid analysis or one of its derivative techniques. The considerations and methods used in the construction and delivery of these sequences differ according to the needs of the assay and the organism chosen as the experimental background.

There are two broad categories of hybrid library: random libraries and cDNA-based libraries. A cDNA library is constituted by the cDNA produced through reverse transcription of mRNA collected from specific cells of types of cell. This library can be ligated into a construct so that it is attached to the BD or AD being used in the assay. A random library uses lengths of DNA of random sequence in place of these cDNA sections. A number of methods exist for the production of these random sequences, including cassette mutagenesis. Regardless of the source of the DNA library, it is ligated into the appropriate place in the relevant plasmid/phagemid using the appropriate restriction endonucleases.

E. Coli-specific Considerations

By placing the hybrid proteins under the control of IPTG-inducible *lac* promoters, they are expressed only on media supplemented with IPTG. Further, by including different antibiotic resistance genes in each genetic construct, the growth of non-transformed cells is easily prevented through culture on media containing the corresponding antibiotics. This is particularly important for counter selection methods in which a *lack* of interaction is needed for cell survival.

The reporter gene may be inserted into the *E. coli* genome by first inserting it into an episome, a type of plasmid with the ability to incorporate itself into the bacterial cell genome with a copy number of approximately one per cell.

The hybrid expression phagemids can be electroporated into *E. coli* XL-1 Blue cells which after amplification and infection with VCS-M13 helper phage, will yield a stock of library phage. These phage will each contain one single-stranded member of the phagemid library.

Recovery of Protein Information

Once the selection has been performed, the primary structure of the proteins which display the appropriate characteristics must be determined. This is achieved by retrieval of the protein-encoding sequences (as originally inserted) from the cells showing the appropriate phenotype.

E. Coli

The phagemid used to transform *E. coli* cells may be "rescued" from the selected cells by infecting them with VCS-M13 helper phage. The resulting phage particles that are produced contain the single-stranded phagemids and are used to infect XL-1 Blue cells. The double-stranded phagemids are subsequently collected from these XL-1 Blue cells, essentially reversing the process used to produce the original library phage. Finally, the DNA sequences are determined through dideoxy sequencing.

Controlling Sensitivity

The *Escherichia coli*-derived Tet-R repressor can be used in line with a conventional reporter gene and can be controlled by tetracycline or doxicycline (Tet-R inhibitors). Thus the expression of Tet-R is controlled by the standard two-hybrid system but the Tet-R in turn controls (represses) the expression of a previously mentioned reporter such as *HIS3*, through its Tet-R promoter. Tetracycline or its derivatives can then be used to regulate the sensitivity of a system utilising Tet-R.

Sensitivity may also be controlled by varying the dependency of the cells on their reporter genes. For example, this may be affected by altering the concentration of histidine in the growth medium for *his3*-dependent cells and altering the concentration of streptomycin for *aadA* dependent cells. Selection-gene-dependency may also be controlled by applying an inhibitor of the selection gene at a suitable concentration. 3-Amino-1,2,4-triazole (3-AT) for example, is a competitive inhibitor of the *HIS3*-gene product and may be used to titrate the minimum level of *HIS3* expression required for growth on histidine-deficient media.

Sensitivity may also be modulated by varying the number of operator sequences in the reporter DNA.

Non-fusion Proteins

A third, non-fusion protein may be co-expressed with two fusion proteins. Depending on the investigation, the third protein may modify one of the fusion proteins or mediate or interfere with their interaction.

Co-expression of the third protein may be necessary for modification or activation of one or both of the fusion proteins. For example, *S. cerevisiae* possesses no endogenous tyrosine kinase. If an investigation involves a protein that requires tyrosine phosphorylation, the kinase must be supplied in the form of a tyrosine kinase gene.

The non-fusion protein may mediate the interaction by binding both fusion proteins simultaneously, as in the case of ligand-dependent receptor dimerization.

For a protein with an interacting partner, its functional homology to other proteins may be assessed by supplying the third protein in non-fusion form, which then may or may not compete with the fusion-protein for its binding partner. Binding between the third protein and the other fusion protein will interrupt the formation of the reporter expression activation complex and thus reduce reporter expression, leading to the distinguishing change in phenotype.

Split-ubiquitin Yeast Two-hybrid

One limitation of classic yeast two-hybrid screens is that they are limited to soluble proteins. It is therefore impossible to use them to study the protein–protein interactions between insoluble integral membrane proteins. The split-ubiquitin system provides a method for overcoming this limitation. In the split-ubiquitin system, two integral membrane proteins to be studied are fused to two different ubiquitin moieties: a C-terminal ubiquitin moiety ("Cub", residues 35–76) and an N-terminal ubiquitin moiety ("Nub", residues 1–34). These fused proteins are called the bait and prey, respectively. In addition to being fused to an integral membrane protein, the Cub moiety is also fused to a transcription factor (TF) that can be cleaved off by ubiquitin specific proteases. Upon bait–prey interaction, Nub and Cub-moieties assemble, reconstituting the split-ubiquitin. The reconstituted split-ubiquitin molecule is recognized by ubiquitin specific proteases, which cleave off the transcription factor, allowing it to induce the transcription of reporter genes.

Fluorescent Two-hybrid Assay

Zolghadr and co-workers presented a fluorescent two-hybrid system that uses two hybrid proteins that are fused to different fluorescent proteins as well as LacI, the lac repressor. The structure of the fusion proteins looks like this: FP2-LacI-bait and FP1-prey where the bait and prey proteins interact and bring the fluorescent proteins (FP1 = GFP, FP2=mCherry) in close proximity at the binding site of the LacI protein in the host cell genome. The system can also be used to screen for inhibitors of protein–protein interactions.

Enzymatic Two-hybrid Systems: KISS

While the original Y2H system used a reconstituted transcription factor, other systems create enzymatic activities to detect PPIs. For instance, the KInase Substrate Sensor ("KISS"), is a mammalian two-hybrid approach has been designed to map intracellular PPIs. Here, a bait protein is fused to a kinase-containing portion of TYK2 and a prey is coupled to a gp130 cytokine receptor fragment. When bait and prey interact, TYK2 phosphorylates STAT3 docking sites on the prey chimera, which ultimately leads to activation of a reporter gene.

One-, Three- and One-two-hybrid Variants

One-hybrid

The one-hybrid variation of this technique is designed to investigate protein–DNA interactions and uses a single fusion protein in which the AD is linked directly to the binding domain. The binding domain in this case however is not necessarily of fixed sequence as in two-hybrid protein–protein analysis but may be constituted by a library. This library can be selected against the desired target sequence, which is inserted in the promoter region of the reporter gene construct. In a positive-selection system, a binding domain that successfully binds the UAS and allows transcription is thus selected.

Note that selection of DNA-binding domains is not necessarily performed using a one-hybrid system, but may also be performed using a two-hybrid system in which the binding domain is varied and the bait and prey proteins are kept constant.

Three-hybrid

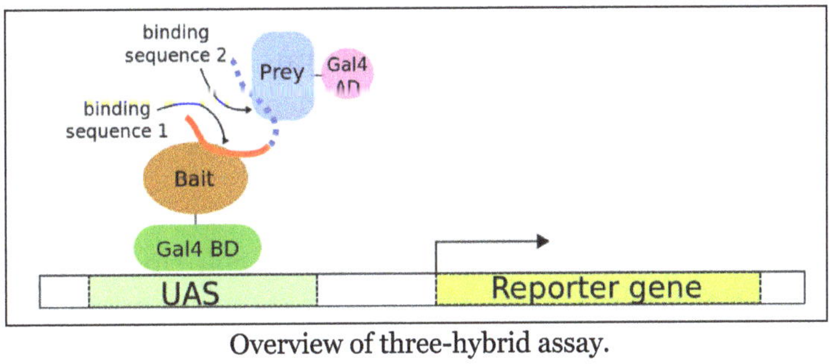

Overview of three-hybrid assay.

RNA-protein interactions have been investigated through a three-hybrid variation of the two-hybrid technique. In this case, a hybrid RNA molecule serves to adjoin together the two protein fusion domains—which are not intended to interact with each other but rather the intermediary RNA molecule (through their RNA-binding domains). Techniques involving non-fusion proteins that perform a similar function, as described in the 'non-fusion proteins' section above, may also be referred to as three-hybrid methods.

One-two-hybrid

Simultaneous use of the one- and two-hybrid methods (that is, simultaneous protein–protein and protein–DNA interaction) is known as a one-two-hybrid approach and expected to increase the stringency of the screen.

Host Organism

Although theoretically, any living cell might be used as the background to a two-hybrid analysis, there are practical considerations that dictate which is chosen. The chosen cell line should be relatively cheap and easy to culture and sufficiently robust to withstand application of the investigative methods and reagents.

Yeast

S. cerevisiae was the model organism used during the two-hybrid technique's inception. It has several characteristics that make it a robust organism to host the interaction, including the ability to form tertiary protein structures, neutral internal pH, enhanced ability to form disulfide bonds and reduced-state glutathione among other cytosolic buffer factors, to maintain a hospitable internal environment. The yeast model can be manipulated through non-molecular techniques and its complete genome sequence is known. Yeast systems are tolerant of diverse culture conditions and harsh chemicals that could not be applied to mammalian tissue cultures.

A number of yeast strains have been created specifically for Y2H screens, e.g. Y187 and AH109, both produced by Clontech. Yeast strains R2HMet and BK100 have also been used.

E. Coli

E. coli-based methods have several characteristics that may make them preferable to yeast-based homologues. The higher transformation efficiency and faster rate of growth lends *E. coli* to the use of larger libraries (in excess of 10^8). A low false positive rate of approximately $3x10^{-8}$, the absence of requirement for a nuclear localisation signal to be included in the protein sequence and the ability to study proteins that would be toxic to yeast may also be major factors to consider when choosing an experimental background organism.

It may be of note that the methylation activity of certain *E. coli* DNA methyltransferase proteins may interfere with some DNA-binding protein selections. If this is anticipated, the use of an *E. coli* strain that is defective for a particular methyltransferase may be an obvious solution.

Applications

Determination of Sequences Crucial for Interaction

the information is totally baseless. By changing specific amino acids by mutating the corresponding DNA base-pairs in the plasmids used, the importance of those amino acid residues in maintaining the interaction can be determined.

After using bacterial cell-based method to select DNA-binding proteins, it is necessary to check the specificity of these domains as there is a limit to the extent to which the bacterial cell genome can act as a sink for domains with an affinity for other sequences (or indeed, a general affinity for DNA).

Drug and Poison Discovery

Protein–protein signalling interactions pose suitable therapeutic targets due to their specificity and pervasiveness. The random drug discovery approach uses compound banks that comprise random chemical structures, and requires a high-throughput method to test these structures in their intended target.

The cell chosen for the investigation can be specifically engineered to mirror the molecular aspect that the investigator intends to study and then used to identify new human or animal therapeutics or anti-pest agents.

Determination of Protein Function

By determination of the interaction partners of unknown proteins, the possible functions of these new proteins may be inferred. This can be done using a single known protein against a library of unknown proteins or conversely, by selecting from a library of known proteins using a single protein of unknown function.

Zinc Finger Protein Selection

To select zinc finger proteins (ZFPs) for protein engineering, methods adapted from the two-hybrid screening technique have been used with success. A ZFP is itself a DNA-binding protein used in the construction of custom DNA-binding domains that bind to a desired DNA sequence.

By using a selection gene with the desired target sequence included in the UAS, and randomising the relevant amino acid sequences to produce a ZFP library, cells that host a DNA-ZFP interaction with the required characteristics can be selected. Each ZFP typically recognises only 3–4 base pairs, so to prevent recognition of sites outside the UAS, the randomised ZFP is engineered into a 'scaffold' consisting of another two ZFPs of constant sequence. The UAS is thus designed to include the target sequence of the constant scaffold in addition to the sequence for which a ZFP is selected.

A number of other DNA-binding domains may also be investigated using this system.

Strengths

- Two-hybrid screens are low-tech; they can be carried out in any lab without sophisticated equipment.
- Two-hybrid screens can provide an important first hint for the identification of interaction partners.
- The assay is scalable, which makes it possible to screen for interactions among many proteins. Furthermore, it can be automated, and by using robots many proteins can be screened against thousands of potentially interacting proteins in a relatively short time.
- Yeast two-hybrid data can be of similar quality to data generated by the alternative approach of coaffinity purification followed by mass spectrometry (AP/MS).

Weaknesses

- The main criticism applied to the yeast two-hybrid screen of protein–protein interactions is the possibility of a high number of false positive (and false negative) identifications. The exact rate of false positive results is not known, but earlier estimates were as high as 70%. The reason for this high error rate lies in the characteristics of the screen.

- Certain assay variants overexpress the fusion proteins which may cause unnatural protein concentrations that lead to unspecific (false) positives.
- The hybrid proteins are fusion proteins; that is, the fused parts may inhibit certain interactions, especially if an interaction takes place at the N-terminus of a test protein (where the DNA-binding or activation domain is typically attached).
- An interaction may not happen in yeast, the typical host organism for Y2H. For instance, if a bacterial protein is tested in yeast, it may lack a chaperone for proper folding that is only present in its bacterial host. Moreover, a mammalian protein is sometimes not correctly modified in yeast (e.g., missing phosphorylation), which can also lead to false results.
- The Y2H takes place in the nucleus. If test proteins are not localized to the nucleus (because they have other localization signals) two interacting proteins may be found to be non-interacting.
- Some proteins might specifically interact when they are co-expressed in the yeast, although in reality they are never present in the same cell at the same time. However, in most cases it cannot be ruled out that such proteins are indeed expressed in certain cells or under certain circumstances.

Each of these points alone can give rise to false results. Due to the combined effects of all error sources yeast two-hybrid have to be interpreted with caution. The probability of generating false positives means that all interactions should be confirmed by a high confidence assay, for example co-immunoprecipitation of the endogenous proteins, which is difficult for large scale protein–protein interaction data. Alternatively, Y2H data can be verified using multiple Y2H variants or bioinformatics techniques. The latter test whether interacting proteins are expressed at the same time, share some common features (such as gene ontology annotations or certain network topologies), have homologous interactions in other species.

Phage Display

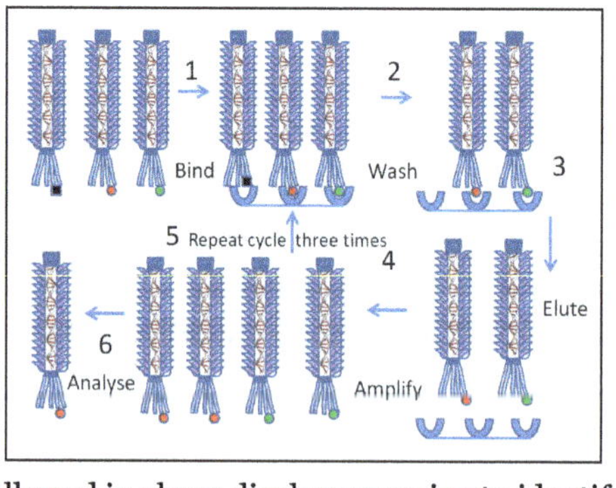

The sequence of events that are followed in phage display screening to identify polypeptides that bind with high affinity to desired target protein or DNA sequence.

Phage display is a laboratory technique for the study of protein–protein, protein–peptide, and protein–DNA interactions that uses bacteriophages (viruses that infect bacteria) to connect proteins with the genetic information that encodes them. In this technique, a gene encoding a protein of interest is inserted into a phage coat protein gene, causing the phage to "display" the protein on its outside while containing the gene for the protein on its inside, resulting in a connection between genotype and phenotype. These displaying phages can then be screened against other proteins, peptides or DNA sequences, in order to detect interaction between the displayed protein and those other molecules. In this way, large libraries of proteins can be screened and amplified in a process called *in vitro* selection, which is analogous to natural selection.

The most common bacteriophages used in phage display are M13 and fd filamentous phage, though T4, T7, and λ phage have also been used.

History

Phage display was first described by George P. Smith in 1985, when he demonstrated the display of peptides on filamentous phage by fusing the peptide of interest onto gene III of filamentous phage. A patent by George Pieczenik claiming priority from 1985 also describes the generation of phage display libraries. This technology was further developed and improved by groups at the Laboratory of Molecular Biology with Greg Winter and John McCafferty, The Scripps Research Institute with Lerner and Barbas and the German Cancer Research Center with Breitling and Dübel for display of proteins such as antibodies for therapeutic protein engineering.

Principle

Like the two hybrid system, phage display is used for the high-throughput screening of protein interactions. In the case of M13 filamentous phage display, the DNA encoding the protein or peptide of interest is ligated into the pIII or pVIII gene, encoding either the minor or major coat protein, respectively. Multiple cloning sites are sometimes used to ensure that the fragments are inserted in all three possible reading frames so that the cDNA fragment is translated in the proper frame. The phage gene and insert DNA hybrid is then inserted (a process known as "transduction") into *Escherichia coli* (E. coli) bacterial cells such as TG1, SS320, ER2738, or XL1-Blue *E. coli*. If a "phagemid" vector is used (a simplified display construct vector) phage particles will not be released from the *E. coli* cells until they are infected with helper phage, which enables packaging of the phage DNA and assembly of the mature virions with the relevant protein fragment as part of their outer coat on either the minor (pIII) or major (pVIII) coat protein. By immobilizing a relevant DNA or protein target(s) to the surface of a microtiter plate well, a phage that displays a protein that binds to one of those targets on its surface will remain while others are removed by washing. Those that remain can be eluted, used to produce more phage (by bacterial infection with helper phage) and so produce a phage mixture that is enriched with relevant (i.e. binding) phage. The repeated cycling of these steps is referred to as 'panning', in reference to the enrichment of a sample of gold by removing undesirable materials. Phage eluted in the final step can be used to infect a suitable bacterial host, from which the phagemids can be collected and the relevant DNA sequence excised and sequenced to identify the relevant, interacting proteins or protein fragments.

The use of a helper phage can be eliminated by using 'bacterial packaging cell line' technology.

Elution can be done combining low-pH elution buffer with sonification, which, in addition to loosening the peptide-target interaction, also serves to detach the target molecule from the immobilization surface. This ultrasound-based method enables single-step selection of a high-affinity peptide.

Applications

Applications of phage display technology include determination of interaction partners of a protein (which would be used as the immobilised phage "bait" with a DNA library consisting of all coding sequences of a cell, tissue or organism) so that the function or the mechanism of the function of that protein may be determined. Phage display is also a widely used method for *in vitro* protein evolution (also called protein engineering). As such, phage display is a useful tool in drug discovery. It is used for finding new ligands (enzyme inhibitors, receptor agonists and antagonists) to target proteins. The technique is also used to determine tumour antigens (for use in diagnosis and therapeutic targeting) and in searching for protein-DNA interactions using specially-constructed DNA libraries with randomised segments.

Competing methods for *in vitro* protein evolution include yeast display, bacterial display, ribosome display, and mRNA display.

Antibody Maturation in Vitro

The invention of antibody phage display revolutionised antibody drug discovery. Initial work was done by laboratories at the MRC Laboratory of Molecular Biology (Greg Winter and John McCafferty), the Scripps Research Institute (Richard Lerner and Carlos F. Barbas) and the German Cancer Research Centre (Frank Breitling and Stefan Dübel). In 1991, The Scripps group reported the first display and selection of human antibodies on phage. This initial study described the rapid isolation of human antibody Fab that bound tetanus toxin and the method was then extended to rapidly clone human anti-HIV-1 antibodies for vaccine design and therapy.

Phage display of antibody libraries has become a powerful method for both studying the immune response as well as a method to rapidly select and evolve human antibodies for therapy. Antibody phage display was later used by Carlos F. Barbas at The Scripps Research Institute to create synthetic human antibody libraries, a principle first patented in 1990 by Breitling and coworkers (Patent CA 2035384), thereby allowing human antibodies to be created in vitro from synthetic diversity elements.

Antibody libraries displaying millions of different antibodies on phage are often used in the pharmaceutical industry to isolate highly specific therapeutic antibody leads, for development into antibody drugs primarily as anti-cancer or anti-inflammatory therapeutics. One of the most successful was HUMIRA (adalimumab), discovered by Cambridge Antibody Technology as D2E7 and developed and marketed by Abbott Laboratories. HUMIRA, an antibody to TNF alpha, was the world's first fully human antibody, which achieved annual sales exceeding $1bn.

General Protocol

Below is the sequence of events that are followed in phage display screening to identify polypeptides that bind with high affinity to desired target protein or DNA sequence:

1. Target proteins or DNA sequences are immobilised to the wells of a microtiter plate.
2. Many genetic sequences are expressed in a bacteriophage library in the form of fusions

with the bacteriophage coat protein, so that they are displayed on the surface of the viral particle. The protein displayed corresponds to the genetic sequence within the phage.

3. This phage display library is added to the dish and after allowing the phage time to bind, the dish is washed.
4. Phage-displaying proteins that interact with the target molecules remain attached to the dish, while all others are washed away.
5. Attached phage may be eluted and used to create more phage by infection of suitable bacterial hosts. The new phage constitutes an enriched mixture, containing considerably less irrelevant phage (i.e. non-binding) than were present in the initial mixture.
6. Steps 3 to 6 are optionally repeated one or more times, further enriching the phage library in binding proteins.
7. Following further bacterial-based amplification, the DNA within in the interacting phage is sequenced to identify the interacting proteins or protein fragments.

Selection of the Coat Protein

Filamentous Phages

pIII

pIII is the protein that determines the infectivity of the virion. pIII is composed of three domains (N1, N2 and CT) connected by glycine-rich linkers. The N2 domain binds to the F pilus during virion infection freeing the N1 domain which then interacts with a TolA protein on the surface of the bacterium. Insertions within this protein are usually added in position 249 (within a linker region between CT and N2), position 198 (within the N2 domain) and at the N-terminus (inserted between the N-terminal secretion sequence and the N-terminus of pIII). However, when using the BamHI site located at position 198 one must be careful of the unpaired Cysteine residue (C201) that could cause problems during phage display if one is using a non-truncated version of pIII.

An advantage of using pIII rather than pVIII is that pIII allows for monovalent display when using a phagemid (Ff-phage derived plasmid) combined with a helper phage. Moreover, pIII allows for the insertion of larger protein sequences (>100 amino acids) and is more tolerant to it than pVIII. However, using pIII as the fusion partner can lead to a decrease in phage infectivity leading to problems such as selection bias caused by difference in phage growth rate or even worse, the phage's inability to infect its host. Loss of phage infectivity can be avoided by using a phagemid plasmid and a helper phage so that the resultant phage contains both wild type and fusion pIII.

cDNA has also been analyzed using pIII via a two complementary leucine zippers system, Direct Interaction Rescue or by adding an 8-10 amino acid linker between the cDNA and pIII at the C-terminus.

pVIII

pVIII is the main coat protein of Ff phages. Peptides are usually fused to the N-terminus of pVIII.

Usually peptides that can be fused to pVIII are 6-8 amino acids long. The size restriction seems to have less to do with structural impediment caused by the added section and more to do with the size exclusion caused by pIV during coat protein export. Since there are around 2700 copies of the protein on a typical phages, it is more likely that the protein of interest will be expressed polyvalently even if a phagemid is used. This makes the use of this protein unfavorable for the discovery of high affinity binding partners.

To overcome the size problem of pVIII, artificial coat proteins have been designed. An example is Weiss and Sidhu's inverted artificial coat protein (ACP) which allows the display of large proteins at the C-terminus. The ACP's could display a protein of 20kDa, however, only at low levels (mostly only monovalently).

pVI

pVI has been widely used for the display of cDNA libraries. The display of cDNA libraries via phage display is an attractive alternative to the yeast-2-hybrid method for the discovery of interacting proteins and peptides due to its high throughput capability. pVI has been used preferentially to pVIII and pIII for the expression of cDNA libraries because one can add the protein of interest to the C-terminus of pVI without greatly affecting pVI's role in phage assembly. This means that the stop codon in the cDNA is no longer an issue. However, phage display of cDNA is always limited by the inability of most prokaryotes in producing post-translational modifications present in eukaryotic cells or by the misfolding of multi-domain proteins.

While pVI has been useful for the analysis of cDNA libraries, pIII and pVIII remain the most utilized coat proteins for phage display.

pVII and pIX

In an experiment in 1995, display of Glutathione S-transferase was attempted on both pVII and pIX and failed. However, phage display of this protein was completed successfully after the addition of a periplasmic signal sequence (pelB or ompA) on the N-terminus. In a recent study, it has been shown that AviTag, FLAG and His could be displayed on pVII without the need of a signal sequence. Then the expression of single chain Fv's (scFv), and single chain T cell receptors (scTCR) were expressed both with and without the signal sequence.

PelB (an amino acid signal sequence that targets the protein to the periplasm where a signal peptidase then cleaves off PelB) improved the phage display level when compared to pVII and pIX fusions without the signal sequence. However, this led to the incorporation of more helper phage genomes rather than phagemid genomes. In all cases, phage display levels were lower than using pIII fusion. However, lower display might be more favorable for the selection of binders due to lower display being closer to true monovalent display. In five out of six occasions, pVII and pIX fusions without pelB was more efficient than pIII fusions in affinity selection assays. The paper even goes on to state that pVII and pIX display platforms may outperform pIII in the long run.

The use of pVII and pIX instead of pIII might also be an advantage because virion rescue may be undertaken without breaking the virion-antigen bond if the pIII used is wild type. Instead, one could cleave in a section between the bead and the antigen to elute. Since the pIII is intact it does not matter whether the antigen remains bound to the phage.

T7 phages

The issue of using Ff phages for phage display is that they require the protein of interest to be translocated across the bacterial inner membrane before they are assembled into the phage. Some proteins cannot undergo this process and therefore cannot be displayed on the surface of Ff phages. In these cases, T7 phage display is used instead. In T7 phage display, the protein to be displayed is attached to the C-terminus of the gene 10 capsid protein of T7.

The disadvantage of using T7 is that the size of the protein that can be expressed on the surface is limited to shorter peptides because large changes to the T7 genome cannot be accommodated like it is in M13 where the phage just makes its coat longer to fit the larger genome within it. However, it can be useful for the production of a large protein library for scFV selection where the scFV is expressed on an M13 phage and the antigens are expressed on the surface of the T7 phage.

Bioinformatics Resources and Tools

Databases and computational tools for mimotopes have been an important part of phage display study. Databases, programs and web servers have been widely used to exclude target-unrelated peptides, characterize small molecules-protein interactions and map protein-protein interactions. Users can use three dimensional structure of a protein and the peptides selected from phage display experiment to map conformational eptiopes. Some of the fast and efficient computational methods are available online (e.g. EpiSearch http://curie.utmb.edu/episearch.html).

Tandem Affinity Purification

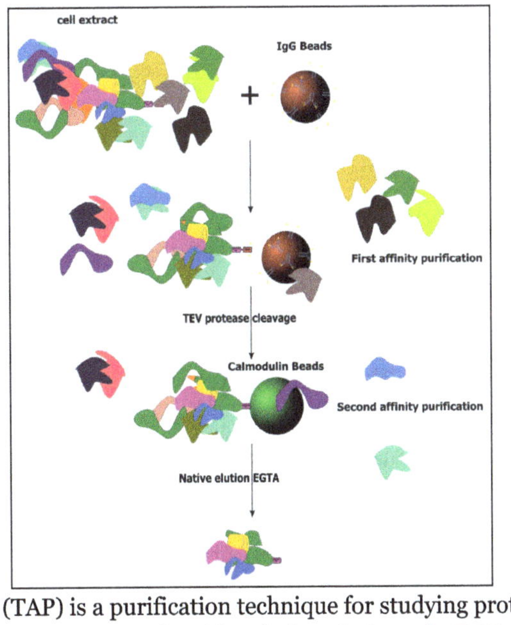

Tandem affinity purification (TAP) is a purification technique for studying protein–protein interactions. It involves creating a fusion protein with a designed piece, the TAP tag, on the end.

In the original version of the technique, the protein of interest with the TAP tag first binds to beads coated with IgG, the TAP tag is then broken apart by an enzyme, and finally a different part of the TAP tag binds reversibly to beads of a different type. After the protein of interest has been washed through two affinity columns, it can be examined for binding partners.

The original TAP method involves the fusion of the TAP tag to the C-terminus of the protein under study. The TAP tag consists of calmodulin binding peptide (CBP) from the N-terminal, followed by tobacco etch virus protease (TEV protease) cleavage site and Protein A, which binds tightly to IgG. The relative order of the modules of the tag is important because Protein A needs to be at the extreme end of the fusion protein so that the entire complex can be retrieved using an IgG matrix.

Many other tag combinations have been proposed since the TAP principle was first published.

Variant Tags

This tag is also known as the C-terminal TAP tag because an N-terminal version is also available. However, the method to be described assumes the use of a C-terminal tag, although the principle behind the method is still the same.

History

TAP tagging was invented by a research team working in the European Molecular Biology Laboratory at late 1990s (Rigaut et al., 1999, Puig et al.,2001) and proposed as a new tool for proteome exploration. It was used by the team to characterize several protein complexes (Rigaut et al., 1999, Caspary et al. 1999, Bouveret et al., 2000, Puig et al., 2001). The first large-scale application of this technique was in 2002, in which the research team worked in collaboration with scientists of the proteomics company Cellzome to develop a visual map of the interaction of more than 230 multi-protein complexes in a yeast cell by systematically tagging the TAP tag to each protein. The first successful report of using TAP tag technology in plants came in 2004 (Rohila et al., 2004,).

Process

There are a few methods in which the fusion protein can be introduced into the host. If the host is yeast, then one of the methods may be the use of plasmids that will eventually translate the fusion protein within the host. Whichever method that is being used, it is preferable to maintain expression of the fusion protein as close as possible to its natural level.

Once the fusion protein is translated within the host, the new protein at one end of the fusion protein would be able to interact with other proteins. Subsequently, the fusion protein is retrieved from the host by breaking the cells and retrieving the fusion protein through affinity selection, together with the other constituents attached to the new protein, by means of an IgG matrix.

After washing, TEV protease is introduced to elute the bound material at the TEV protease cleavage site. This eluate is then incubated with calmodulin-coated beads in the presence of calcium. This second affinity step is required to remove the TEV protease as well as traces of contaminants remaining after the first affinity step. After washing, the eluate is then released with ethylene glycol tetraacetic acid (EGTA).

The native elution, consisting of the new protein and its interacting protein partners as well as CBP, can now be analyzed by sodium dodecyl sulfate polyacrylamide gel electrophoresis (SDS-PAGE) or be identified by mass spectrometry.

Advantages

An advantage of this method is that there can be real determination of protein partners quantitatively in vivo without prior knowledge of complex composition. It is also simple to execute and often provides high yield. One of the obstacles of studying protein protein interaction is the contamination of the target protein especially when we don't have any prior knowledge of it. TAP offers an effective, and highly specific means to purify target protein. After 2 successive affinity purifications, the chance for contaminants to be retained in the eluate reduces significantly.

Disadvantages

However, there is also the possibility that a tag added to a protein might obscure binding of the new protein to its interacting partners. In addition, the tag may also affect protein expression levels. On the other hand, the tag may also not be sufficiently exposed to the affinity beads, hence skewing the results.

There may also be a possibility of a cleavage of the proteins by the TEV protease, although this is unlikely to be frequent given the high specificity of the TEV protease.

Suitability

As this method involves at least 2 rounds of washing, it may not be suitable for screening transient protein interactions, unlike the yeast two-hybrid method or *in vivo* crosslinking with photo-reactive amino acid analogs. However, it is a good method for testing stable protein interactions and allows various degrees of investigation by controlling the number of times the protein complex is purified.

Applications

In 2002, the TAP tag was first used with mass spectrometry in a large-scale approach to systematically analyse the proteomics of yeast by characterizing multiprotein complexes. The study revealed 491 complexes, 257 of them wholly new. The rest were familiar from other research, but now virtually all of them were found to have new components. They drew up a map relating all the protein components functionally in a complex network.

Many other proteomic analyses also involve the use of TAP tag. A research by EMBO (Dziembowski, 2004) identified a new complex required for nuclear pre-mRNA retention and splicing. They have purified a novel trimeric complex composed of 3 other subunits (Snu17p, Bud13p and Pml1p) and find that these subunits are not essential for viability but required for efficient splicing (removal of introns) of pre-mRNA. In 2006, *Fleischer et al.* systematically identified proteins associated with eukaryotic ribosomal complexes. They used multifaceted mass spectrometry proteomic screens to identify yeast ribosomal complexes and then used TAP tagging to functionally link up all these proteins.

Other Epitope-tag Combinations

The principle of tandem-affinity purification of multiprotein complexes is not limited to the combination of CBP and Protein A tags used in the original work by Rigaut et al. (1999). For example, the combination of FLAG- and HA-tags has been used since 2000 by the group of Nakatani to purify numerous protein complexes from mammalian cells. Many other tag combinations have been proposed since the TAP principle was published.

Bimolecular Fluorescence Complementation

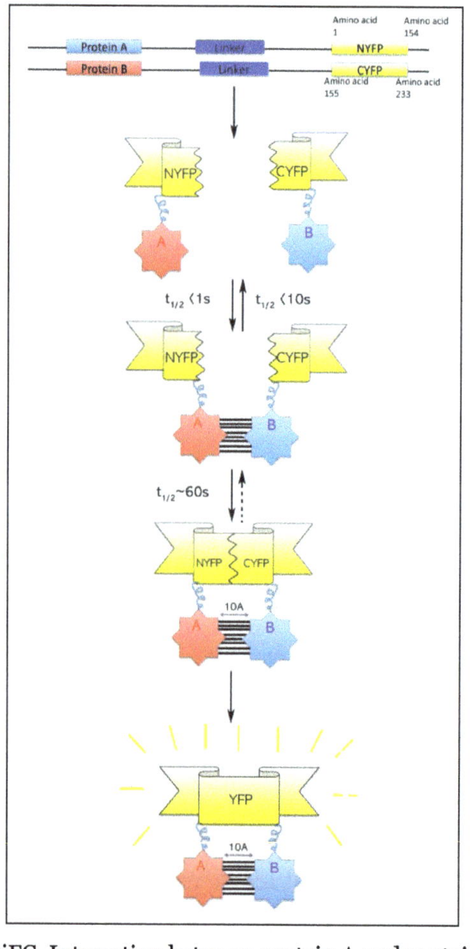

Protein complex formation using BiFC. Interaction between protein A and protein B occurs first, followed by the reformation and fluorescence of fluorescent reporter protein.

Bimolecular fluorescence complementation (also known as BiFC) is a technology typically used to validate protein interactions. It is based on the association of fluorescent protein fragments that are attached to components of the same macromolecular complex. Proteins that are postulated to interact are fused to unfolded complementary fragments of a fluorescent reporter protein and expressed in live cells. Interaction of these proteins will bring the fluorescent fragments within proximity, allowing the reporter protein to reform in its native three-dimensional structure and emit its fluorescent signal. This fluorescent signal can be detected and located within the cell using an inverted fluorescence microscope that allows imaging of fluorescence in cells. In addition, the intensity of the fluorescence emitted is proportional to the strength of the interaction, with stronger levels of fluorescence indicating close or direct interactions and lower fluorescence levels suggesting interaction within a complex. Therefore, through the visualisation and analysis of the intensity and distribution of fluorescence in these cells, one can identify both the location and interaction partners of proteins of interest.

History

Biochemical complementation was first reported in subtilisin-cleaved bovine pancreatic ribonuclease, then expanded using β-galactosidase mutants that allowed cells to grow on lactose.

Recognition of many proteins' ability to spontaneously assemble into functional complexes as well as the ability of protein fragments to assemble as a consequence of the spontaneous functional complex assembly of interaction partners to which they are fused was later reported for ubiquitin fragments in yeast protein interactions.

In 2000, Ghosh *et al* developed a system that allowed a green fluorescent protein (GFP) to be reassembled using an anti-parallel leucine zipper in *E. coli* cells. This was achieved by dissecting GFP into C- and N-terminal GFP fragments. As the GFP fragment was attached to each leucine zipper by a linker, the heterodimerisation of the anti-parallel leucine zipper resulted in a reconstituted, or re-formed, GFP protein that could be visualised. The successful fluorescent signal indicated that the separate GFP peptide fragments were able to correctly reassemble and achieve tertiary folding. It was, therefore, postulated that using this technique, fragmented GFP could be used to study interaction of protein–protein pairs that have their N–C termini in close proximity.

After the demonstration of successful fluorescent protein fragment reconstitution in mammalian cells, Hu *et al.* described the use of fragmented yellow fluorescent protein (YFP) in the investigation of bZIP and Rel family transcription factor interactions. This was the first report bZIP protein interaction regulation by regions outside of the bZIP domain, regulation of subnuclear localization of the bZIP domains Fos and Jun by their different interacting partners, and modulation of transcriptional activation of bZIP and Rel proteins through mutual interactions. In addition, this study was the first report of an *in vivo* technique, now known as the bimolecular fluorescence complementation (BiFC) assay, to provide insight into the structural basis of protein complex formation through detection of fluorescence caused by the assembly of fluorescent reporter protein fragments tethered to interacting proteins.

Fluorescent Labeling

Fluorophore activation occurs through an autocatalytic cyclization reaction that occurs after the protein has been folded correctly. This was advanced with the successful reconstitution of the YFP fluorophore from protein fragments that had been fused to interacting proteins within 8 hours of transfection was reported in 2002.

Workflow

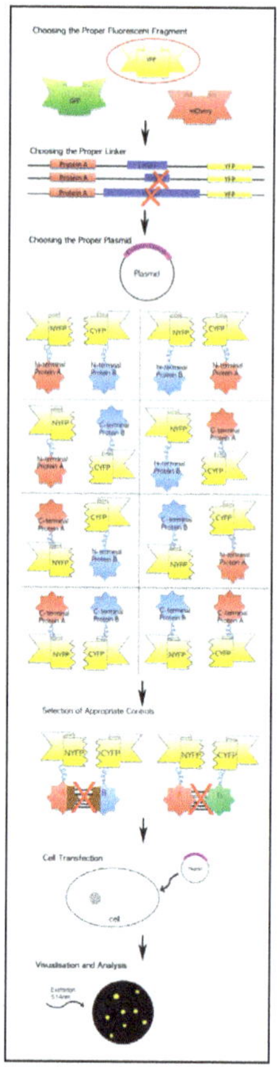

Wo4rkflow for BiFC.

Selection of Fusion Protein Production System

There are different production systems that can be used for the fusion protein generated. Transient gene expression is used to identify protein–protein interactions *in vivo* as well as in subcellular localisation of the BiFC complex. However, one must be cautious against protein over-expression, as this may skew both preferential localisation and the predominant protein complexes formed. Instead, weak promoters, the use of low levels of plasmid DNA in the transfection, and plasmid vectors that do not replicate in mammalian cells should be used to express proteins at or near their endogenous levels to mimic the physiological cellular environment. Also, careful selection of the fluorescent protein is important, as different fluorescent proteins require different cellular environments. For example, GFP can be used in *E. coli* cells, while YFP is used in mammalian cells.

Stable cell lines with the expression vector integrated into its genome allows more stable gene expression in the cell population, resulting in more consistent results.

Determination of Fusion Sites

When deciding the linker fusion site on the protein surface, there are three main considerations. First, the fluorescent protein fragments must be able to associate with one another when their tethered proteins interact. Structural information and the location of the interaction surface may be useful when determining the fusion site to the linker, although the information is not necessary, as multiple combinations and permutations can be screened. Secondly, the creation of the fusion protein must not significantly alter the localisation, stability, or expression of the proteins to which the fragments are linked as compared to the endogenous wild-type proteins. Finally, the addition of the fluorescent fragment fusion must not affect the biological function of the protein, preferably verified using assays that evaluate all of the proteins' known functions.

Designing Linkers

A linker is a short amino acid sequence that tethers the fluorescent reporter protein fragment to the protein of interest, forming the fusion protein. When designing a linker sequence, one must ensure that the linker is sufficiently soluble and long to provide the fluorescent protein fragments with flexibility and freedom of movement so that the fragment and its partner fragment will collide frequently enough to reconstitute during the interaction of their respective fused proteins. Although it is not documented, it is possible that the length or the sequence of the linker may influence complementation of some proteins. Reported linker sequences RSIAT and RPACKIPNDLKQKVMNH (single amino acid code) and AAANSSIDLISVPVDSR (Sigma) have been successfully used in BiFC experiments.

Creating Proper Plasmid Expression Vectors

When designing plasmid vectors to express the proteins of interest, the construct must be able to express proteins that are able to form fusion proteins with fluorescent protein fragments without disrupting the protein's function. In addition, the expected protein complex must be able to accept stabilisation of the fluorescent protein fragment interaction without affecting the protein complex function or the cell being studied. Many fluorescent protein fragments that combine in several ways can be used in BiFC. Generally, YFP is recommended to serve as the reporter protein, cleaved at residue 155 (N-terminal consisting of residues 1–154 and C-terminal consisting of residues 155–238) or residue 173 in particular, as these sets of fragments are highly efficient in their complementation when fused to many interacting proteins and they produce low levels fluorescence when fused to non-interacting proteins. It is suggested that each target protein is fused to both the N- and C-terminal fragments of the fluorescent reporter protein in turn, and that the fragments are fused at each of the N- and C-terminal ends of the target proteins. This will allow a total of eight different permutations, with interactions being tested:

N-terminal fragment fused at the N-terminal protein 1 + C-terminal fragment fused at the N-terminal protein 2N-terminal fragment fused at the N-terminal protein 1 + C-terminal fragment fused at the C-terminal protein 2N-terminal fragment fused at the C-terminal protein 1 + C-terminal fragment fused at the N-terminal protein 2N-terminal fragment fused at the C-terminal protein 1 + C-terminal fragment fused at the C-terminal protein 2C-terminal fragment fused at the N-terminal protein 1 + N-terminal fragment fused at the N-terminal protein 2C-terminal fragment fused at the N-terminal protein 1 + N-terminal fragment fused at the C-terminal protein 2C-terminal fragment fused at the C-terminal protein 1 + N-terminal fragment fused at the N-terminal protein 2C-terminal fragment fused at the C-terminal protein 1 + N-terminal fragment fused at the C-terminal protein 2.

Selection of Appropriate Cell Culture System

As previously stated, it is important to ensure that the fluorescent reporter protein being used in

BiFC is appropriate and can be expressed in the cell culture system of choice, as not all reporter proteins can fluoresce or be visualised in all model systems.

Selection of Appropriate Controls

Fluorescent protein fragments can associate and fluoresce at low efficiency in the absence of a specific interaction. Therefore, it is important to include controls to ensure that the fluorescence from fluorescent reporter protein reconstitution is not due to unspecific contact.

Some controls include fluorophore fragments linked to non-interacting proteins, as the presence of these fusions tend to decrease non-specific complementation and false positive results.

Another control is created by linking the fluorescent protein fragment to proteins with mutated interaction faces. So long as the fluorescent fragment is fused to the mutated proteins in the same manner as the wild-type protein, and the gene expression levels and localisation are unaffected by the mutation, this serves as a strong negative control, as the mutant proteins, and therefore, the fluorescent fragments, should be unable to interact.

Internal controls are also necessary to normalise for differences in transfection efficiencies and gene expression in different cells. This is accomplished by co-transfecting cells with plasmids encoding the fusion proteins of interest as well as a whole (non-fragmented) protein that fluoresces at a different wavelength from the fluorescent reporter protein. During visualisation, one determines the fluorescence intensities of the BiFC complex and the internal control which, after subtracting background signal, becomes a ratio. This ratio represents the BiFC efficiency and can be compared with other ratios to determine the relative efficiencies of the formation of different complexes.

Cell Transfection

Once the fusion proteins and controls have been designed and generated in their appropriate expression system, the plasmids must be transfected into the cells to be studied. After transfection, one must wait, typically about eight hours, to allow time for the fusion proteins to interact and their linked fluorescent reporter protein fragments to associate and fluoresce.

Visualisation and Analysis

After sufficient time for the fusion proteins and their linked fluorescent fragments to interact and fluoresce, the cells can be observed under an inverted fluorescence microscope that can visualise fluorescence in cells. Although the fluorescence intensity of BiFC complexes is usually <10% of that produced by expression of intact fluorescent proteins, the extremely low autofluorescence in the visible range extremely most cells often makes the BiFC signal orders of magnitude higher than background fluorescence.

If fluorescence is detected when the fusion proteins are expressed, but is lacking or significantly reduced after the expression of the mutated negative control, it is likely that a specific interaction occurs between the two target proteins of interest. However, if the fluorescence intensity is not significantly different between the mutated negative control fusion protein and its wild-type counterpart, then the fluorescence is likely caused by non-specific protein interactions, so a different combination of fusion protein conformations should be tested.

If no fluorescence is detected, an interaction may still exist between the proteins of interest, as the creation of the fusion protein may alter the structure or interaction face of the target protein or the fluorescence fragments may be physically unable to associate. To ensure that this result is not a false negative, that there is no interaction, the protein interaction must be tested in a situation where fluorescence complementation and activation requires an external signal. In this case, if the external signal fails to cause fluorescence fragment association, it is likely that the proteins do not interact or there is a physical impediment to fluorescence complementation.

Strengths

Relevant Biological Context

Proteins interact with different protein partners and other macromolecules to achieve functions that support different functions in cells that support survival of the organism. Identifying these interactions may provide clues to their effects on cell processes. As these interactions can be affected by both the internal environment and external stimuli, studying these interactions *in vivo* and at endogenous levels, as is recommended in BiFC, provides a physiologically-relevant context from which to draw conclusions about protein interactions.

Direct Visualisation

BiFC enables direct visualisation of protein interactions in living cells with limited cell perturbation, rather than relying on secondary effects or staining by exogenous molecules that can fail to distribute evenly. This, and the ability to observe the living cells for long periods of time, is made possible by the strong intrinsic fluorescence of the reconstituted reporter protein reduces the chances of an incorrect readout associated with the protein isolation process.

Sensitivity

Unlike many *in vivo* protein-interaction assays, BiFC does not require protein complexes to be formed by a large proportion of the proteins or at stoichiometric proportions. Instead, BiFC can detect interactions among protein subpopulations, weak interactions, and low expression proteins due to the stable complementation of the fluorescent reporter protein. In addition, successful fluorescent protein reconstitution has been reported for protein partners over 7 nm apart, so long as the linkers binding the fluorophore fragment to the protein of interest has the flexibility needed to associate with its corresponding fragment. Furthermore, the strength of the protein interaction can be quantitatively determined by changes in fluorescent signal strength.

Spatial Resolution

BiFC allows measurement of spatial and temporal changes in protein complexes, even in response to activating and inhibiting drugs and subcellularly, providing the highest spatial resolution of *in vivo* protein–protein interaction assays.

No Specialised Equipment

BiFC does not require specialised equipment, as visualisation is possible with an inverted fluorescence microscope that can detect fluorescence in cells. In addition, analysis does not require complex data processing or correction for other sources of fluorescence.

No structural Information Needed

BiFC can be performed without structural information about the interaction partners, so long as the fluorescent reporter protein fragments can associate within the complex, as multiple combinations of fusion proteins can be screened. This is due to the assumption that, since the protein functions are recapitulated in the *in vivo* context, the complex structure will resemble that of the intact proteins seen physiologically.

Multiple Applications

The BiFC technology has been refined and expanded to include the abilities to simultaneously visualise multiple protein complexes in the same cell, RNA/protein interactions, to quickly detect changes in gene transduction pathways, demonstrate hidden phenotypes of drugs, where the predicted treatment outcome (i.e. cell death, differentiation, morphological change) is not seen *in vivo*, study complex formation in different cellular compartments, and to map protein interaction surfaces.

Limitations

Real-time Detection

The fluorescent signal only is produced after the proteins have interacted, which is generally in the order of hours. Hence BiFC is unable to provide real-time detection of protein interactions. The delay for chemical reactions to generate fluorophore may also have an effect on the dynamics of complex dissociation and partner exchange.

Irreversible BiFC Formation

BiFC complex formation is only reversible during the initial step of fluorescent reporter protein re-assembly, typically in the order of milliseconds. Once the fluorochrome has been reconstituted, it is essentially irreversible *in vitro*. This prevents proteins from interacting with others and may disrupt the association/disassociation of protein complexes in dynamic equilibrium.

Independent Fluorescent Protein Fragment Associations

Fluorescent protein fragments have a limited ability to associate independent of the proteins to which they are fused. Although protein-independent association will vary depending on identities of the fusion proteins and their expression levels, one must provide the necessary and numerous controls to distinguish between true and false-positive protein interactions. Generally, this limitation is mitigated by ensuring that the fusion proteins of interest are expressed at endogenous concentrations.

Altering Protein Structure and Steric Hindrance

Fluorescent fragment linkage may alter the folding or structure of the protein of interest, leading to the elimination of an interacting protein's surface binding site. In addition, the arrangement of the fluorescent fragments may prevent fluorophore reconstitution through steric hindrance, although steric hindrance can be reduced or eliminated by using a linker sequence that allows sufficient flexibility for the fluorescent fragments to associate. Therefore, absence of fluorescence complementation may be a false negative and does not necessarily prove that the interaction in question does not occur.

Obligate Anaerobes

Due to the requirement of molecular oxygen for fluorophore formation, BiFC cannot be used in obligate anaerobes, which cannot survive in the presence of oxygen. This limits the use of BiFC to aerobic organisms.

Use of Fusion Proteins

Because endogenous wild-type proteins cannot be visualised *in vivo*, fusion proteins must be created and their plasmids transfected into the cells studied. These fusion proteins may not recapitulate the functions, localisation, and interactions common to their wild-type counterparts, providing an inaccurate picture of the proteins in question. This problem can be alleviated by using structural information and the location of interaction sites to rationally identify fusion sites on the proteins of interest, using appropriate controls, and comparing the expression levels and functions of the fusion and wild-type proteins through Western Blots and functional assays.

Temperature Dependence

Although low temperatures favour the reconstitution of fluorescence when fragments are within proximity, this may affect the behaviour of the target proteins leading to inaccurate conclusions regarding the nature of protein interactions and their interacting partners.

Exact Interaction Relationship Unknown

Because fluorophore reconstitution can occur at a distance of 7 nm or more, fluorescence complementation may indicate either a direct or indirect (i.e. within the same complex) interaction between the fluorescent fragments' fused proteins.

Application

In addition to the validation of protein–protein interactions described above, BiFC has been expanded and adapted to other applications:

Multicolour Fluorescence

The fluorescent protein fragments used in BiFC have been expanded to include the colours blue, cyan, green, yellow, red, cherry, and Venus. This range in colours has made the development of multicolour fluorescence complementation analysis possible. This technique allows multiple protein complexes to be visualised simultaneously in the same cell. In addition, proteins typically have a large number of alternative interaction partners. Therefore, by fusing fragments of different fluorescent proteins to candidate proteins, one can study competition between alternative interaction partners for complex formation through the complementation of different fluorescent colour fragments.

RNA-binding Protein Interactions

BiFC has been expanded to include the study of RNA-binding protein interactions in a method Rackham and Brown described as trimolecular fluorescence complementation (TriFC). In this method, a fragment of the Venus fluorescent protein is fused to the mRNA of interest, and the complementary Venus portion fused to the RNA-binding protein of interest. Similar to BiFC, if the mRNA and protein interact, the Venus protein will be reconstituted and fluoresce. Also known as the RNA bridge method, as the fluorophore and other interacting proteins form a bridge between the protein and the RNA of interest, this allows a simple detection and localisation of RNA-protein interactions within a living cell and provides a simple method to detect direct or indirect RNA-protein association (i.e. within a complex) that can be verified through in vitro analysis of purified compounds or RNAi knockdown of the bridging molecule(s).

Pathway Organisation and Signal Transduction Cascades

BiFC can be used to link genes to one other and their function through measurement of interactions among the proteins that the genes encode. This application is ideal for novel genes in which little is known about their up- and downstream effectors, as novel pathway linkages can be made. In addition, the effects of drugs, hormones, or deletion or knockdown of the gene of interest, and the subsequent effects on both the strength of the protein–protein interactions and the location of the interaction can be observed within seconds.

Complex Formation in Different Cellular Compartments

BiFC has been used to study nuclear translocation, via complex localisation, as well as interactions involving integral membrane proteins. Thus, BiFC is an important tool in understanding transcription factor localisation in subcellular compartments.

Quantifying Protein–protein Interaction Surfaces

BiFC has been coupled with flow cytometry (BiFC-FC). This allows protein–protein interaction surfaces to be mapped through the introduction of site-directed or random mutations that affect complex formation.

Comparisons to Other Technologies

Most techniques used to study protein–protein interactions rely on *in vitro* methods. Unfortunately, studying proteins in an artificial system, outside of their cellular environment, poses a number of difficulties. For example, this may require the removal of proteins from their normal cellular environment. The processing required to isolate the protein may affect its interactions with other proteins. In addition, isolating the protein from the intracellular signalling and mechanisms that occur in the normal cell may provide a misleading picture of intracellular and physiological occurrences. Furthermore, proteins studied in vitro may be studied at concentrations vastly different from their normal abundance levels, may not necessarily be transported efficiently into the cells, or may not be selective enough to function in the host genome. Finally, by studying proteins *in vitro*, one is unable to determine the influence of specific protein–protein interactions in the cell on the functional or physiological consequences.

Other *in vivo* assays most commonly used to study protein–protein interactions include fluorescence resonance energy transfer (FRET) and yeast two-hybrid (Y2H) assay. Each of these assays has their advantages and disadvantages in comparison to BiFC:

Fluorescence Resonance Energy Transfer (FRET)

Fluorescence resonance energy transfer (FRET), also known as förster resonance energy transfer, resonance energy transfer (RET) or electronic energy transfer (EET), is based on the transfer of energy from an excited (donor) chromophore or fluorophore (if the chromophores are fluorescent) to a nearby acceptor. In this method, fluorophores are chemically linked or genetically fused to two proteins hypothesised to interact. If the proteins interact, this will bring the fluorophores into close spatial proximity. If the fluorophores are oriented in a manner that exposes the fluorophores to one another, usually ensured when designing and constructing the fluorophore-protein linkage/fusion, then the energy transfer from the excited donor fluorophore will result in a change in the fluorescent intensities or lifetimes of the fluorophores.

Yeast Two-hybrid (Y2H)

The yeast two-hybrid (Y2H) is a genetic screening technique that can be used to detect physical (binding) protein–protein or protein–DNA interactions. It tests a 'bait' protein of known function that is fused to the binding domain of the transcription factor GAL4 against potential interacting proteins or a cDNA library that express the GAL4 activation domain (the 'prey').

Technology Comparisons

Comparison Technology	Similarity to BiFC	Advantages	Disadvantages	
FRET	Ability to detect and locate protein interaction sites within live cells	**Instantaneous real-time monitoring of protein interactions** This allows the detection of proteins with short-lived associations. **Reversible fluorophore interaction** More complex interaction dynamics can be detected (i.e. dynamic equilibrium – continuous complex formation and dissociation).	**Close spatial proximity** Potential interacting proteins must be close to one another, typically within 60–100Å, for the energy transfer to occur. In contrast, reconstitution of the BiFC fluorophore is possible at a distance of over 7nm and only requires that the link between the fluorophore fragment and the protein of interest be sufficiently flexible to allow association. **Decreased sensitivity** BiFC is typically more sensitive due to its stable protein complementation. In contrast, FRET produces background fluorescence when the acceptor fluorophore is excited. Therefore, numerous controls must be performed to quantify the change in fluorescence intensity in the presence versus the absence of energy transfer between fluorophores. Consequently, it is difficult to detect weak interactions using FRET . FRET usually requires higher levels of gene expression to detect the energy transfer, as the proportion of proteins that form complexes must be high enough to produce a detectable change in donor and acceptor fluorescence intensities. **Irreversible photo-bleaching** FRET quantification requires irreversible photobleaching or fluorescence lifetime imaging that uses instrumentation that is not widely available. This complicates the visualisation process, as the fluorescence will be destroyed over time by the light needed to excite the fluorophores.	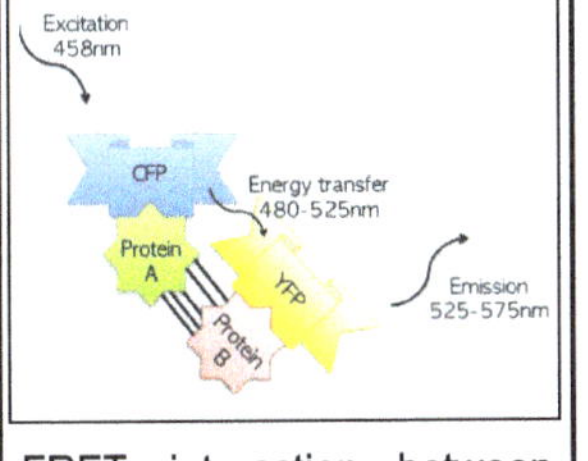 FRET: interaction between protein A and protein B brings the two fluorescent proteins together and energy transfer occurs between the two fluorescent proteins

Y2H	*In vivo* technique used to screen for interactions	**Genetic interaction screen** Y2H provides an *in vivo* technique for studying transcription factor binding at the sequence level in eukaryotic cells	**Tentative bait-prey linkage** The link between the bait and prey is often tentative and often ambiguous, as the study of over-expressed fusion proteins in the yeast nucleus can give rise to non-specific interactions, and incompatible systems (i.e. mammalian proteins do not always express correctly in yeast cells, proteins may co-express unnaturally). **Erroneous transcription activation** The reporter gene can easily be transcriptionally activated, providing a false positive readout of protein–protein or protein–DNA interaction. **Genetic complementation** Different alleles of the same gene can result in genetic complementation, leading to inaccurate protein–DNA interactions. **Overexpression of proteins** Y2H requires proteins to be overexpressed, which can skew data analysis, as proteins may interact with different proteins or DNA when expressed at higher-than-normal levels as compared to the interactions that occur at endogenous gene expression levels. **Nuclear localisation** Y2H can only be used in the nucleus, where transcriptional activation occurs. This both removes the protein from its physiologically relevant environment, disregarding the role of the protein's normal environment on protein interactions, and limits detection to proteins that are active in the nucleus.	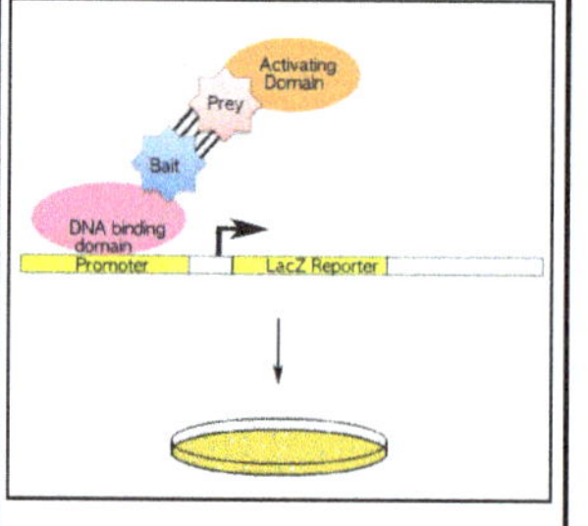 Yeast-2-Hybrid: interaction between protein A and protein B activates transcription

Chromatin Immunoprecipitation

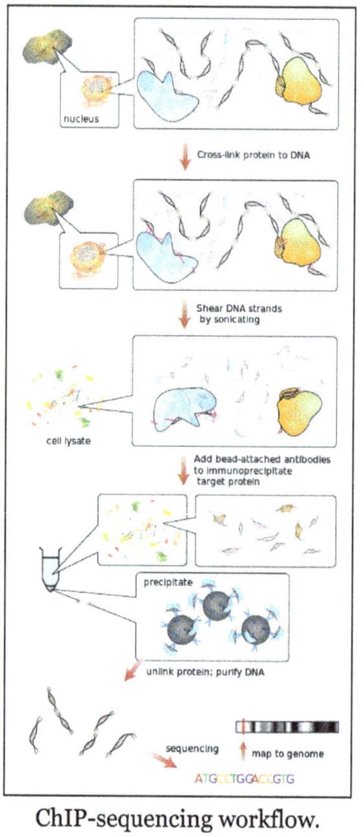

ChIP-sequencing workflow.

Chromatin Immunoprecipitation (ChIP) is a type of immunoprecipitation experimental technique used to investigate the interaction between proteins and DNA in the cell. It aims to determine whether specific proteins are associated with specific genomic regions, such as transcription factors on promoters or other DNA binding sites, and possibly defining cistromes. ChIP also aims to determine the specific location in the genome that various histone modifications are associated with, indicating the target of the histone modifiers.

Briefly, the conventional method is as follows:

1. DNA and associated proteins on chromatin in living cells or tissues are crosslinked (this step is omitted in Native ChIP).
2. The DNA-protein complexes (chromatin-protein) are then sheared into ~500 bp DNA fragments by sonication or nuclease digestion.
3. Cross-linked DNA fragments associated with the protein(s) of interest are selectively immunoprecipitated from the cell debris using an appropriate protein-specific antibody.
4. The associated DNA fragments are purified and their sequence is determined. Enrichment of specific DNA sequences represents regions on the genome that the protein of interest is associated with *in vivo*.

Typical ChIP

There are mainly two types of ChIP, primarily differing in the starting chromatin preparation. The first uses reversibly cross-linked chromatin sheared by sonication called cross-linked ChIP (XChIP). Native ChIP (NChIP) uses native chromatin sheared by micrococcal nuclease digestion.

Cross-linked ChIP (XChIP)

Cross-linked ChIP is mainly suited for mapping the DNA target of transcription factors or other chromatin-associated proteins, and uses reversibly cross-linked chromatin as starting material. The agent for reversible cross-linking could be formaldehyde or UV light. Then the cross-linked chromatin is usually sheared by sonication, providing fragments of 300 - 1000 base pairs (bp) in length. Mild formaldehyde crosslinking followed by nuclease digestion has been used to shear the chromatin. Chromatin fragments of 400 - 500bp have proven to be suitable for ChIP assays as they cover two to three nucleosomes.

Cell debris in the sheared lysate is then cleared by sedimentation and protein–DNA complexes are selectively immunoprecipitated using specific antibodies to the protein(s) of interest. The antibodies are commonly coupled to agarose, sepharose or magnetic beads. The immunoprecipitated complexes (i.e., the bead–antibody–protein–target DNA sequence complex) are then collected and washed to remove non-specifically bound chromatin, the protein–DNA cross-link is reversed and proteins are removed by digestion with proteinase K. An epitope-tagged version of the protein of interest, or *in vivo* biotinylation can be used instead of antibodies to the native protein of interest.

The DNA associated with the complex is then purified and identified by polymerase chain reaction

(PCR), microarrays (ChIP-on-chip), molecular cloning and sequencing, or direct high-throughput sequencing (ChIP-Seq).

Native ChIP (NChIP)

Native ChIP is mainly suited for mapping the DNA target of histone modifiers. Generally, native chromatin is used as starting chromatin. As histones wrap around DNA to form nucleosomes, they are naturally linked. Then the chromatin is sheared by micrococcal nuclease digestion, which cuts DNA at the length of the linker, leaving nucleosomes intact and providing DNA fragments of one nucleosome (200bp) to five nucleosomes (1000bp) in length.

Thereafter, methods similar to XChIP are used for clearing the cell debris, immunoprecipitating the protein of interest, removing protein from the immunoprecipated complex, and purifying and analyzing the complex-associated DNA.

Comparison of XChIP and NChIP

The major advantage for NChIP is antibody specificity. It is important to note that most antibodies to modified histones are raised against unfixed, synthetic peptide antigens and that the epitopes they need to recognize in the XChIP may be disrupted or destroyed by formaldehyde cross-linking, particularly as the cross-links are likely to involve lysine e-amino groups in the N-terminals, disrupting the epitopes. This is likely to explain the consistently low efficiency of XChIP protocols compare to NChIP.

But XChIP and NChIP have different aims and advantages relative to each other. XChIP is for mapping target sites of transcription factors and other chromatin associated proteins; NChIP is for mapping target sites of histone modifiers.

Table 1 Advantages and Disadvantages of NChIP and XChIP

	XChIP	NChIP
Advantages	Suitable for transcriptional factors, or any other weakly binding chromatin associated proteins. Applicable to any organisms where native protein is hard to prepare	Testable antibody specificity Better antibody specificity as target protein naturally intact Better chromatin and protein recovery efficiency due to Better antibody specificity
Disadvantages	Inefficient chromatin recovery due to antibody target protein epitope disruption May cause false positive result due to fixation of transient proteins to chromatin Wide range of chromatin shearing size due to random cut by sonication.	Usually not suitable for non-histone proteins Nucleosomes may rearrange during digestion

History and New ChIP Methods

In 1984 John T. Lis and David Gilmour, at the time a graduate student in the Lis lab, used UV irradiation, a zero-length protein-nucleic acid crosslinking agent, to covalently cross-link proteins bound to DNA in living bacterial cells. Following lysis of cross-linked cells and immunoprecipitation of bacterial RNA polymerase, DNA associated with enriched RNA polymerase was hybridized to probes corresponding to different regions of known genes to determine the in vivo distribution and density of RNA polymerase at these genes. A year later they used the same methodology to study distribution of eukaryotic RNA polymerase II on fruit fly heat shock genes. These reports are considered the pioneering studies in the field of chromatin immunoprecipitation. XChIP was further modified and developed by Alexander Varshavsky and co-workers, who examined distribution of histone H4 on heat shock genes using formaldehyde cross-linking. This technique was extensively developed and refined thereafter. NChIP approach was first described by Hebbes *et al.*, 1988, and also been developed and refined quickly. The typical ChIP assay usually take 4–5 days, and require 10^6~ 10^7 cells at least. Now new techniques on ChIP could be achieved as few as 100~1000 cells and complete within one day:

- Carrier ChIP (CChIP): This approach could use as few as 100 cells by adding *Drosophila* cells as carrier chromatin to reduce loss and facilitate precipitation of the target chromatin. However, it demands highly specific primers for detection of the target cell chromatin from the foreign carrier chromatin background, and it takes two to three days.

- Fast ChIP (qChIP): The fast ChIP assay reduced the time by shortening two steps in a typical ChIP assay: *(i)* an ultrasonic bath accelerates the rate of antibody binding to target proteins—and thereby reduces immunoprecipitation time *(ii)* a resin-based (Chelex-100) DNA isolation procedure reduces the time of cross-link reversal and DNA isolation. However, the fast protocol is suitable only for large cell samples (in the range of 10^6~10^7). Up to 24 sheared chromatin samples can be processed to yield PCR-ready DNA in 5 hours, allowing multiple chromatin factors be probed simultaneously and/or looking at genomic events over several time points.

- Quick and quantitative ChIP (Q^2ChIP) : The assay uses 100,000 cells as starting material and is suitable for up to 1,000 histone ChIPs or 100 transcription factor ChIPs. Thus many chromatin samples can be prepared in parallel and stored, and Q^2ChIP can be undertaken in a day.

- MicroChIP (μChIP): chromatin is usually prepared from 1,000 cells and up to 8 ChIPs can be done in parallel without carriers. The assay can also start with 100 cells, but only suit for one ChIP. It can also use small (1 mm^3) tissue biopsies and microChIP can be done within one day.

- Matrix ChIP: This is a microplate-based ChIP assay with increased throughput and simplified the procedure. All steps are done in microplate wells without sample transfers, enabling a potential for automation. It enables 96 ChIP assays for histone and various DNA-bound proteins in a single day.

ChIP has also been applied for genome wide analysis by combining with microarray technology (ChIP-on-chip) or second generation DNA-sequencing technology (Chip-Sequencing). ChIP can also combine with paired-end tags sequencing in Chromatin Interaction Analysis using Paired End Tag sequencing (ChIA-PET), a technique developed for large-scale, de novo analysis of higher-order chromatin structures.

Limitations of ChIP

- Large Scale assays using ChIP is challenging using intact model organisms. This is because antibodies have to be generated for each TF, or, alternatively, transgenic model organisms expressing epitope-tagged TFs need to be produced.
- Researchers studying differential gene expression patterns in small organisms also face problems as genes expressed at low levels, in a small number of cells, in narrow time window.
- ChIP experiments cannot discriminate between different TF isoforms (Protein isoform).

References

- Kevin Downard (24 August 2007). Mass Spectrometry of Protein Interactions. John Wiley & Sons. ISBN 978-0-470-14632-3. Retrieved 17, February 2020
- "DNA Microarrays: Techniques". Arabidopsis.info. Archived from the original on August 28, 2008. Retrieved 07, June 2020
- Lowman HB, Clackson T (2004). "1.3". Phage display: a practical approach. Oxford [Oxfordshire]: Oxford University Press. pp. 10–11. ISBN 0-19-963873-X
- Scott JS, Barbas CF III, Burton, DA (2001). Phage Display: A Laboratory Manual. Plainview, N.Y: Cold Spring Harbor Laboratory Press. ISBN 0-87969-740-7

Permissions

All chapters in this book are published with permission under the Creative Commons Attribution Share Alike License or equivalent. Every chapter published in this book has been scrutinized by our experts. Their significance has been extensively debated. The topics covered herein carry significant information for a comprehensive understanding. They may even be implemented as practical applications or may be referred to as a beginning point for further studies.

We would like to thank the editorial team for lending their expertise to make the book truly unique. They have played a crucial role in the development of this book. Without their invaluable contributions this book wouldn't have been possible. They have made vital efforts to compile up to date information on the varied aspects of this subject to make this book a valuable addition to the collection of many professionals and students.

This book was conceptualized with the vision of imparting up-to-date and integrated information in this field. To ensure the same, a matchless editorial board was set up. Every individual on the board went through rigorous rounds of assessment to prove their worth. After which they invested a large part of their time researching and compiling the most relevant data for our readers.

The editorial board has been involved in producing this book since its inception. They have spent rigorous hours researching and exploring the diverse topics which have resulted in the successful publishing of this book. They have passed on their knowledge of decades through this book. To expedite this challenging task, the publisher supported the team at every step. A small team of assistant editors was also appointed to further simplify the editing procedure and attain best results for the readers.

Apart from the editorial board, the designing team has also invested a significant amount of their time in understanding the subject and creating the most relevant covers. They scrutinized every image to scout for the most suitable representation of the subject and create an appropriate cover for the book.

The publishing team has been an ardent support to the editorial, designing and production team. Their endless efforts to recruit the best for this project, has resulted in the accomplishment of this book. They are a veteran in the field of academics and their pool of knowledge is as vast as their experience in printing. Their expertise and guidance has proved useful at every step. Their uncompromising quality standards have made this book an exceptional effort. Their encouragement from time to time has been an inspiration for everyone.

The publisher and the editorial board hope that this book will prove to be a valuable piece of knowledge for students, practitioners and scholars across the globe.

Index

www.ingramcontent.com/pod-product-compliance
Lightning Source LLC
LaVergne TN
LVHW080726210926
848485LV00003B/525